W0225665

The Economics of Blockchain Consensus

Joshua Gans

The Economics of Blockchain Consensus

Exploring the Key Tradeoffs in Blockchain Design

Joshua Gans
Rotman School of Management
University of Toronto
Toronto, ON, Canada

ISBN 978-3-031-33082-7   ISBN 978-3-031-33083-4   (eBook)
https://doi.org/10.1007/978-3-031-33083-4

This Palgrave Macmillan imprint is published by the registered company Springer Nature
Switzerland AG
The registered company address is: Gewerbestrasse 11, 6330 Cham, Switzerland

ACKNOWLEDGEMENTS

Special thanks to Eric Budish, Ethan Buchman, Agostino Capponi, Scott Kominers Scott Stornetta and Richard Titus for helpful discussions. I owe a great debt to my co-authors on blockchain research: Christian Catalini, Neil Gandal, Guillaume Haeringer, Hanna Halaburda and Richard Holden. I also want to acknowledge Tim Roughgarden, whose video series on the Foundations of Blockchain helped me navigate a dense literature.

CONTENTS

LIST OF FIGURES

CHAPTER 1

Introduction

Abstract This chapter discusses the intertwined origins of economics and computer science, highlighting the invention of the blockchain by the mysterious Satoshi Nakamoto. The chapter emphasises the importance of Nakamoto's innovation in creating a distributed ledger system, the permissionless blockchain, that requires no trust relationships. The chapter also delves into the economic implications of cryptocurrencies, arguing that tokens are essential to the operation of decentralised systems. The intention is to explore the inner workings of blockchain consensus and to make this literature more accessible to economists. The book focuses on trade-offs, such as security versus speed, and permissioned versus permissionless networks, and examines the incentives behind Proof of Work and Proof of Stake blockchains.

Keywords Economics · Computer science · Blockchain · Nakamoto · Consensus

For as long as there has been economics and computer science, there has been a relationship between the two disciplines. Charles Babbage was a leading economist who developed a theory of the division of cognitive labour (Babbage, 1832) before turning to invent the first

© The Author(s), under exclusive license to Springer Nature Switzerland AG 2023
J. Gans, *The Economics of Blockchain Consensus*,
https://doi.org/10.1007/978-3-031-33083-4_1

computer, the Analytical Engine. John von Neumann famously developed the theory of games (Von Neumann & Morgenstern, 1944) and contributed to growth theory before developing the hardware/software design for modern computers. Herbert Simon won a Nobel prize in economics for his theory of bounded rationality and a Turing Award in computer science for his advances in developing artificial intelligence. So perhaps it should be no surprise that the computer science of operating distributed networks should receive a significant advance from the same person (the still unknown Satoshi Nakamoto [2008]) who cracked the problem of how to launch a digital currency. But just over a decade and a half ago, that happened, and most were surprised.

The common origins of economics and computer science have diverged over the years. This is why Nakamoto's innovation was so surprising. The challenge was to build a distributed network that did not require pre-registration or trust in the nodes that operated it. Nakamoto showed how to take three separate computer science innovations—the blockchain, cryptography and the notion of Proof of Work—and combine them to construct a permissionless blockchain, a distributed ledger that required no trust relationships to reach a consensus as to what the contents of the ledger were.

But what motivated Nakamoto was not an advance in computer science but instead an advance in monetary economics. Money had gone digital some decades ago. While cash-on-hand is a key part of the economy, most of the ways we use money are by sending messages from one bank to another to debit and credit different accounts accordingly. The banks then keep their own records along with a series of regulations to ensure they do not take advantage of their position to create too much money out of thin air. For Nakamoto, however, there was concern about the privileged position of the banks themselves. And in 2008, there was ample reason not to trust them as holders of the monetary system as the world was in the midst of its biggest financial crisis since the Great Depression of the 1930s.

Nakamoto had realised that it was the record-keeping part of the monetary economy that needed to be preserved but that no clear and anointed institution should be the one preserving it. Indeed, anyone, anywhere, with a computer could play their part. By embedding some rules in code that could not be changed without a broad consensus, Nakamoto showed that digital tokens (which were names bitcoins) could be created and allocated to different users securely in cryptographically locked wallets. The wallets were just entries on a ledger. But importantly, those with tokens registered to them would be the only ones able to move those tokens from their own wallet to someone else on the registry. In economics, Narayana Kocherlakota (1998) had demonstrated that money could be represented as entries in a ledger in a way that could operate as a memory to account for productive activity in the economy. Kocherlakota was talking theoretically. Bitcoin, when it emerged, was that theory becoming a reality.

Economists have struggled to understand how seemingly valuable digital tokens could just be created by code and, at the very least, come to behave like financial assets that have some, at least purported, relationship to the real economy. But it is safe to say that deep down, economists always knew money could just come into being; in particular, Keynes (1937) and Fama (1980). The task of understanding cryptocurrencies is far from done in economics. But it is important to recognise that Nakamoto's contribution was unprecedented. In just nine pages (well, eight plus references), an important branch of both fields was revolutionised.

I am an economist who has been interested in these developments; although not in any way that would have allowed me to obtain some share of the riches as cryptocurrency went from nothing to something. But the weeds of Nakamoto's innovation have intrigued me. Setting aside the monetary aspects of all of this, the idea that it would be possible to create sustainable and distributed ledgers of anything under the guise of what we now term 'blockchain technologies' seemed to offer the potential for streamlining so many institutions; in particular, those around contracting (Catalini & Gans, 2020; Gans, 2022). This is perhaps the attitude of what many like to claim their interest in these technologies are—'blockchain, not bitcoin'—but I have come to believe that the token aspects developed by Nakamoto and others cannot be separated from the vision of a decentralised ledger that might be applied to matters beyond money. Instead, the tokens themselves and their exchangeability with the 'real world' (Gans & Halaburda, 2015) play a key role in financing the operation of decentralised systems. If you want those systems to be permissionless, there is no separating them.

Thus, arises this book. My intention is to steer away from the monetary and financial aspects of cryptocurrencies. Instead, I want to look at the inner workings of the blockchain consensus that is at the heart of it all. Those developments have, to date, largely lay with computer science. That is natural. That is where the code that runs all of this comes from. But at the same time, as I explored those workings more closely, there was an undercurrent of economics going on. The very notion of Proof of Work was both an entry fee into a game and also a way of creating incentives for players in that game to operate in the interests of the whole. The notion of Proof of Stake was similar, except that this game could be fully bootstrapped and seemingly did not require the use of real resources at all. Computer scientists have long recognised the relationship (Halpern & Moses, 1990, for example). Economists had paid little attention.

One of the reasons for that, which I found when starting to research in this area, was that the computer science literature, for all its overlap with game theory and potentially rational agents, was quite impenetrable to economists like myself. Part of this was language, but the more critical barrier was in the nature of theorems and proofs computer scientists would use as well as their choice of what criteria were important in motivating particular designs of things like blockchain protocols. With some persistence, I started to get a better feel for the differences. And in doing so, I came to the realisation that maybe writing a book focusing on blockchain consensus could reduce the barriers to others. It could translate computer science results using the language of game theory but also provide assurances ('You're not crazy, these choices of what to care about are weird') and identify gaps ('Yes, understanding incentives more closely might yield a better design'). I expect that computer scientists might, therefore, find the chapters that follow baffling in their own choices. But that is not my audience. I wanted to transpose my struggles into a form that could open up that literature to more economists.

To take one example. Computer scientists designing blockchain protocols seem to shy away from marginal thinking. For instance, computer scientists often lay out desired technical properties of network protocols, such as speed in reaching an outcome and/or economising on communication or computational costs. Economists, too, are concerned with technical properties, but rather than stating a technical metric, they like to consider what the value of a little more speed or a little extra reduction in computational costs will generate for network users. Simply put, economists want to specify the full payoff from different designs, whereas computer scientists often take for granted that more of some technical property is desirable. Thus, in this book on blockchain consensus, I begin in Chapter 2 by considering the value of that consensus. That requires understanding what blockchains do, some of their technical limitations but also the potential role that blockchains might play in providing expanded contractual solutions as a potential substitute for other trust mechanisms. The simple message is that economists will care about the demand for services blockchains might provide as well as the supply conditions for those blockchains.

One consequence of the technical approach from computer scientists is that in the blockchain literature, at least, they are often focused on taking a given protocol and examining whether it achieves certain desired technical specifications. Interestingly, it is often the case that it is not possible to achieve those specifications. Thus, as we will explore in Chapter 3, computer science is littered with various impossibility theorems—for example, showing that it is not possible for a distributed network to achieve perfect security at the fastest possible speed. This often creates, in their mind, a choice between one or the other. Thus, on a graph such as Fig. 1.1, computer scientists often frame the choice as between points X and Y with the notion that if you care more about security than the speed, you will choose X and vice versa. Instead, in reality, the technical trade-off is there but not as stark. You can increase security by a lot if you are starting from a point where you have lots of speed and much less if you have a slow protocol. What is more likely in this instance is that you will want a protocol that is neither perfectly secure nor as fast as possible but something intermediate like Z. Where Z is will depend on the marginal rate of substitution between security and speed in the utility function of whoever is designing the protocol.

The outline of this book is structured to focus on those trade-offs and lay them out in a way that is 'economics friendly.' Chapter 3 deals with the trade-off between speed and security, which is the hallmark of studies of distributed networks, of which blockchains are a specific example. Chapter 4 then turns to consider the choice of whether a blockchain is permissioned or permissionless. Prior to Nakamoto, all distributed ledgers were analysed as being permissioned where the nodes operating the network were known, and the biggest unknown was whether they were faulty or not. Permissionless networks, on the other hand, allowed any agent to operate a node and, thus, needed explicit incentives to ensure that they were somewhat secure and operated somewhat a timely manner. It is there that we will examine the operation of Proof of Work blockchains like Bitcoin.

Fig. 1.1 Trade-offs
between security and
speed

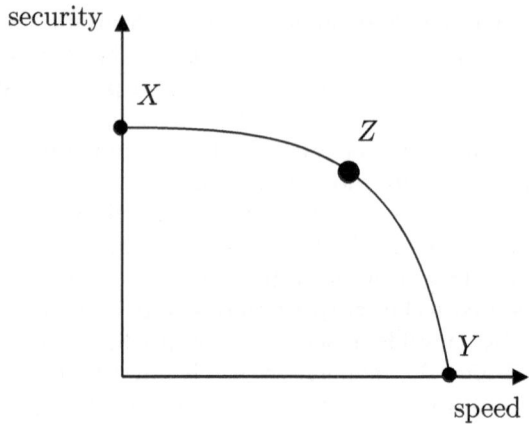

Chapter 5 then compares the differing incentive structures for security generated by Proof of Work and Proof of Stake blockchains. While it may be surprising to some that this trade-off comes late in the book, given the amount of discussion it has engendered with respect to Bitcoin, in particular, and Ethereum's switch in 2022, it is here because of its place in the history of computer science thought. In the chapter, it is demonstrated that the incentives in Proof of Work are paid for upfront and partly by each transaction, while the incentives in Proof of Stake are largely ex post. That potentially makes Proof of Stake networks inherently more efficient at the margin than Proof of Work.

Chapter 6 then examines the role of cryptography in keeping network operations safe. Cryptography hides key information from the view of others, but that also makes it difficult or impossible for those others to create incentives for good behaviour. That trade-off is outlined by taking a close look at a challenge facing blockchains today; that of front-running.

Finally, Chapter 7 examines whether economics and, notably, game theory/mechanism design can play a role in actually improving the operation of blockchain networks. It is argued that for Proof of Stake networks, there is an opportunity to code mechanisms into the protocol that automatically remove tokens from agents with the threat of such removal motivating better behaviour. This is an understudied area where computer science and economics can meet, and an argument is made for more research investment in this area.

REFERENCES

Babbage, C. (1832). *On the economy of machinery and manufactures*. Taylor & Francis.

Catalini, C., & Gans, J. S. (2020). Some simple economics of the blockchain. *Communications of the ACM, 63*(7), 80–90.

Fama, E. F. (1980). Banking in the theory of finance. *Journal of Monetary Economics, 6*(1), 39–57.

Gans, J. S. (2022). The fine print in smart contracts. In M. Corrales Compagnucci, M. Fenwick, & S. Wrbka (Eds.), *Smart contracts technological, business and legal perspectives* (Chapter 2). Hart Publishing.

Gans, J. S., & Halaburda, H. (2015). Some economics of private digital currency. *Economic Analysis of the Digital Economy*, 257–276.

Halpern, J. Y., & Moses, Y. (1990). Knowledge and common knowledge in a distributed environment. *Journal of the ACM (JACM), 37*(3), 549–587.

Keynes, J. M. (1937). The general theory of employment. *The Quarterly Journal of Economics, 51*(2), 209–223.

Kocherlakota, N. R. (1998). Money is memory. *Journal of Economic Theory, 81*(2), 232–251.

Nakamoto, S. (2008). Bitcoin: A peer-to-peer electronic cash system. *Decentralized Business Review*, 21260.

Von Neumann, J., & Morgenstern, O. (1944). *Theory of games and economic behavior*. Princeton University Press.

CHAPTER 2

The Value of Blockchain Consensus

Abstract Blockchains are a unique type of distributed ledger, maintained by multiple nodes that store data and update it through appended
adjustments. These systems enable reliable time-stamping of digital documents and assets, opening up new possibilities for verification. Blockchain
consensus ensures that any node can provide information on database
entries and records, leading to cost reductions and automated contract
verification, including cryptocurrency holdings. By offering cheaper verification methods, blockchain technology allows for contracts to be
enforced in environments where traditional trust mechanisms, like relationships and reputation, would typically be necessary.

Keywords Blockchains · Distributed ledgers · Time-stamping ·
Consensus · Verification

We begin not by considering how blockchain consensus is achieved but
by what we get when it can be done. This capture will explore the value
of blockchain consensus. In short, when a ledger is distributed, the value
of consensus is that any user can query the ledger at any one of the nodes
housing it and be assured that the information retrieved is the same at all
nodes. This can, in turn, foster alternative foundations for trust, reliability

and verification, which, combined, can enrich the space of contracts that can be struck between parties.

In conducting this evaluation, we will take the opportunity to define blockchains carefully as well as examine how trust emerges when blockchains are not available. After all, the value of new technology is assessed relative to what it substitutes for.

2.1   Defining Blockchains

A blockchain is a type of distributed ledger. To understand this, let's build-up from the ledger to distributed and then to the blockchain.

Ledgers

A ledger is a record of what we will call here *transactions*. Transactions are entries into the ledger that update the ledger's state. For instance, suppose there are three agents—'Ariel' (A), 'Bailey' (B) and 'Casey' (C)[1]—who each have an initial endowment, w_i ($i \in \{A, B, C\}$), of tokens. A transaction may involve Ariel giving Bailey 3 tokens recorded as $(3, A, B)$. That is, a transaction is a tuple (x, S, R), where x is the amount sent, S is the identity of the sender, and R is the identity of the receiver. The state of the ledger then becomes $(w_A - 3, w_B + 3, w_C)$ (see Fig. 2.1).

Of course, a ledger need not be a record of tokens but could be a record of other things (e.g., assets more generally or identifiers), or it could be a data repository of some kind. In that case, a transaction is

Address	Entry
Ariel	w_A
Bailey	w_B
Casey	w_C

$M_t = (3, A, B)$
$\longrightarrow$

Address	Entry
Ariel	$w_A - 3$
Bailey	$w_B + 3$
Casey	w_C

Fig. 2.1   Updating the ledger

---

[1] In computer science, these agents are usually Alice, Bob and Charlie, but I have opted for gender-free variants.

a modifier to a particular entry. For instance, an insurance contract may use the ledger to work out if a farmer should receive a payout for rainfall insurance. The ledger would record the amount of rainfall in a location over a specified time period, being periodically updated with additional rainfall data. The contract would then use the final state of the ledger (or a particular entry) to trigger whether a payment to the farmer should be made or not.

Distributed Ledgers

Ledgers have existed throughout civilsation. Typically, there is the main ledger, and copies may be kept as a backup or for use elsewhere. Thus, while copies are a form of distribution, ledgers kept with respect to a single main ledger are called non-distributed.

By contrast, a distributed ledger is one where multiple copies of the ledger are kept with multiple parties, which we will refer to as *nodes*. Let's start with a situation where all nodes have the same copy of the ledger. Then a transaction is received by one or more nodes. This would cause them to update their copy of the ledger. By implication, unless the transaction is received by all nodes, at that moment, there would be no consensus amongst the nodes as to what the true ledger is. Thus, there would have to be a protocol by which consensus could be achieved. For instance, any nodes receiving a transaction could resend it to all other nodes. In principle, with a little time, consensus could be achieved again.

There are, of course, challenges that mean this 'in principle' solution would not work well in practice. We will examine those throughout this book. For the moment, it should be noted that, at least in principle, just because ledgers are distributed does not preclude them being relied upon as references by agents so long as consensus is reached.

Distributed ledgers offer security and reliability benefits over their non-distributed counterparts. A non-distributed ledger has a single point of failure—the main ledger—that can be hacked or could be subject to communications difficulties. By contrast, when a ledger is distributed, a hack would require taking control of a larger number of nodes simultaneously and if any one node had communication difficulties, the other nodes would still be available.

This is the reason why computer scientists often relate distributed ledgers and the problem of achieving consensus to the more general problem of state machine replication (or SMR). The security issues that confronted stock exchanges and air traffic controllers in the 1980s motivated research in the area. That research, in turn, formed the basis of research into achieving consensus in a distributed ledger and then a blockchain system.

Blocks of Transactions

A distributed ledger can be any sort of database that is distributed, updated and synchronised amongst multiple nodes. A blockchain is a type of distributed ledger that updates itself in a very specific fashion. Transactions are appended to a chain in blocks. For instance, Bitcoin blocks can contain anything from 0 to 4000 transactions depending on how many have occurred during a 10-minute interval.

Importantly, blocks are time-stamped and point to the immediate predecessor (or parent) block. Thus, if block B_t is added at time t, it contains, along with transactions, a pointer to a block, B_{t-1} and the next subsequent block, B_{t+1} points to B_t. The sequence of blocks, $\{B_\tau\}_{\tau=0}^t$, can be traced back all the way to B_0, which is called the *genesis block* and which, by definition, contains the initial state of the ledger and no pointer to a previous block (see Fig. 2.2 for an example).

Thus, a blockchain involves a series of instructions that is appended only to the genesis block's ledger. This makes the blockchain less than a ledger itself but an audit trail of changes to a ledger. Importantly, this means that, except for an attack or very short-run latency issues, no past transactions in a blockchain can be lost. By contrast, in a ledger, entries can be rewritten, and there is not necessarily an account of the changes. The blockchain builds up from that account and is designed to never lose the record of changes so long as it continues to operate.

Having an audit trail is a very useful feature. Importantly, it offers the potential to avoid the costs associated with auditing that arise in many situations, most importantly, the keeping of financial accounts. Auditing

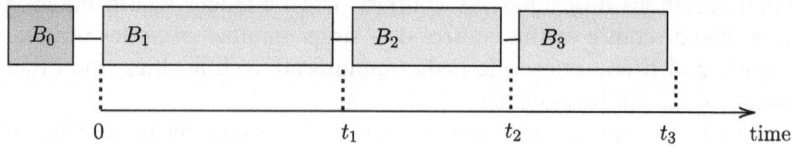

Fig. 2.2 A blockchain

is a process of tracing how particular entries in a ledger or database came to be. Accounting rules are made to make auditing easier, but there are still costs associated with checking and double-checking. And that is only if nothing problematic has occurred.

By contrast, a blockchain builds in an audit trail and enforces it as the ledger is updated. This obviates the need for an ex post audit entirely. For instance, suppose that A gives B 3 tokens (as before) at time t. Then at a subsequent time $> t$, B gives C 2 of those tokens. This can rule out certain disputes. For instance, A might claim those 2 tokens were not B's to transfer. However, the fact that a transaction where A had transferred those tokens to B already existed resolves that dispute. More critically, a blockchain can provide a way of time-stamping any digital document— from token ownership to anything else. It is this feature that gives the blockchain an opportunity to have broad value by reducing transaction (or, more generally, verification) costs.[2]

How to Timestamp a Digital Document

Tokens are a form of digital document. In blockchains, tokens each have unique identifiers. For instance, a bitcoin has an identifier (like the serial number you find on cash notes) and can itself be broken into smaller units. The smallest unit, complete with its own identifier, is called a satoshi.

The origin of the notion of a blockchain came from two computer scientists, Stuart Haber and Scott Stornetta (1990), who were examining the problem of whether a digital document could be time-stamped without having to keep a full copy of the document in a database. This would be necessary if someone were too confident that the document had not been changed since the time it was stamped. Bad actors could have incentives to back or forward-date a document in order to absolve themselves from obligations. At the time Haber and Stornetta were researching, time-stamping of non-digital documents usually involved certification and being held by a trusted third party which was costly and would obviate the advantages of having digital documents. The problem was that digital documents were potentially easier to tamper with. Moreover, because of privacy concerns, storage costs and the risk of corrupted

---

[2] Catalini and Gans (2020) provides a full discussion of how the blockchain can reduce or eliminate the costs of verification.

files, Haber and Stornetta believed it would be naive to rely on a third party in such environments even if they could be trusted (which was also uncertain).

The first step was to take a digital document and construct a 'hash' of it. A hash (see Box: What is a Hash?) is a unique string of characters that is produced by taking a digital document (or digital anything) and running it through a hash function. The string created can only (or at least almost certainly) be created by the document that has originated it. Thus, even a document of any size (say a mega or gigabyte) will generate a hash of relatively few (say 256) characters. This means that all you would have to have a record of was the hash. Then, in the future, if you had a digital document that claimed to be the same document, you could run it through the hash function, and it should generate the same hash. If not, it is different in some way. That such one-way hash functions exist is a marvel of mathematical knowledge. What is more, such one-way hash functions could also produce digital signatures that uniquely identify the signer. (See Box: What is a Digital Signature?) Finally, the date and time can be appended to the document. What is hashed is the document, signatures (if any) and the time. You would need all three to verify the document and time are the same ones today as were generated in the past.

While this was an elegant solution to the privacy, storage and corruption issues, the holder of the hash would still have to be trusted. Haber and Stornetta proposed that this problem could be resolved if, when the document is first hashed, the hash was stored with other documents in a block, and then each block hashed and pointed to the previous block. In other words, they proposed a blockchain.

The advantage of using a blockchain to time-stamp documents is that, if a document were time-stamped, it would be bundled together with many other documents. Therefore, if someone, including the third party holding the blockchain, wanted to tamper with a document, changing its date would require having it on hand, re-signing and changing many other documents. To be sure, recreating all or almost all of the chain could allow an alteration, but this would require considerable effort.[3]

---

[3] It is this feature of a blockchain that allows for the creation of digitally unique assets. Tokens, whether fungible or not, cannot be duplicated in the sense of being assigned to more than one entry in the current state of the blockchain. Time-stamping is a more general ledger that associates uniqueness with time as well as other factors.

To deal with even this possibility, Haber and Stornetta proposed distributing the hash associated with each block of documents. Their eventual solution was to publish the hash of documents weekly in the *New York Times*' lost and found section. The idea is that the millions of copies of the *Times* were a public record that even those running a repository could not tamper with. They set up a company to actually provide this service, and it has been in operation for over three decades.

What is a Hash? A hash is a procedure by which a digital asset is converted into a string of bits of a specific length. Mathematically, hash functions are a family of functions $h : \{0, 1\}^* \rightarrow \{0, 1\}^l$ that compresses bit-strings of an arbitrary length, $*$, to ones of a fixed length, l, satisfying the properties (1) that the functions, h, are easy to compute and it is easy to select a member of the family at random; and (2) it is not computationally feasible, given one of these functions h, to find a pair of distinct strings that collide; i.e., you cannot find x, x' so that $h(x) = h(x')$. These hash functions are now readily available. For instance, the hash (Sha-256) of the first two sentences of this box would be 7603F2F278417CA94A 1699DEFAA1FE46DD089CE630E4328EBA34CF638BBDB43B. Change any part of that sentence (say, the first use of 'hash' to 'Hash'), and the hash becomes E087B504792CDC4B083 DACC5DBC0CE17D277B797587B719E3AF9226844CD2EB2. If you possessed the hash, you could verify whether the text it was claimed to hash had been altered or not.

What is a Digital Signature? A digital signature that is used in blockchains comes from cryptography. If a sender is sending a receiver an encrypted message, the receiver provides the sender with their public key. The sender uses the public key to encrypt the message. That message is then sent directly to the receiver, or it could be broadcast publicly. Regardless, the only way it can be decrypted and read is by use of the receiver's private key that is associated with the public key already sent through. Thus, so long as the receiver keeps their private key private, no one else can decrypt the message encrypted using the receiver's public key. In this way, it can be verified that someone has signed a particular message or transaction.

Distributed Blockchain Networks

The description of the blockchain thus far does not require more than one node to operate. However, that one node could, in principle, alter the entire chain of transactions and hence, change the audit trail. For this reason, most blockchains are developed to be distributed amongst multiple nodes. This removes the problem that there might otherwise be a single point of failure.

Interestingly, nodes that participate in administering a blockchain divide themselves into two classes. First, there are *full nodes* that hold a complete copy of the blockchain going all the way back to the genesis block (or perhaps some more recent checkpoint). Second, there are validating or *light nodes* that participate by accumulating transactions to form into blocks and then assist in communicating those blocks to other nodes. These nodes need only have available the information required to confirm a new block. When blockchains also have a virtual machine associated with them, e.g., for executing code in smart contracts, some nodes will also have to perform those execution functions if called upon by a transaction.

2.2   THE DRIVER OF VALUE

A blockchain is a technology that builds in an audit trail so that any entry into a ledger has a timestamp of when it was created and when and how it has been altered. If it works as intended, that audit trail is immutable (i.e., you can be assured that it has not been altered). What makes that technology valuable?

Trust

Consider a very simple contracting environment. Suppose that A is supplying a service to B that costs A c to provide and generates a value of v to B. We assume that $v > c$, so there are gains to trade. They agree on a price, p ($< v$ and $> c$). With this price, they sign a contract where A is obliged to provide a service to B for which B will pay A, p once the service is complete. As A anticipates getting paid p upon rendering the service, A's expected profit is $p - c$ while B ends up with a surplus of $v - p$, which in each case is better than receiving 0.

For this transaction to work as intended, A has to rely on B actually paying once the service has been provided. But what if B claims that they agreed to a price of $p - \Delta$, which is less than p? A then has a shortfall of Δ in profits. Figure 2.3 shows the (extensive form) game played between A and B. You will notice that B always has an incentive to abuse A's trust and to push for a lower price. If $p - \Delta - c > 0$, even anticipating this, A will choose to supply the service to B. But Δ is a choice variable for B. B can't force A to 'pay' to supply the service (even after the fact), but they can set $\Delta = p$ and receive the service without paying for it at all. In this case, A would anticipate receiving $-c$ and so would choose not to trust B for a payment at all. A would not supply the service, and both they and B would earn 0. Without something to compel B to pay A after the service is supplied, A would not trust B, and so this mutually beneficial transaction would not take place.[4]

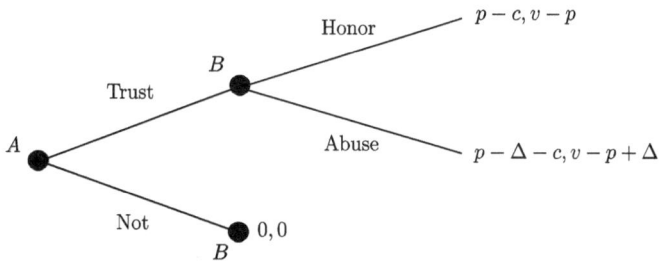

Fig. 2.3 Contracting game between A and B

---

[4] It is possible that B is a trustworthy person and would never consider abusing A's trust. In that situation, the challenge is whether A believes that B is trustworthy with a sufficient likelihood.

What might restore trust here? One option would be for A to be paid *before* supplying the service to B. In that case, however, it is easy to see that once they receive a payment, A has no incentive to expend the costs c in providing the service. In other words, B would not trust A just as A would not trust B.

Other options become possible when the transaction is not a 'once off.' For instance, if A is continually supplying B with a service, then B may not want to avoid paying because they would lose the value from the *relationship* with A. Similarly, even if B did not expect to deal with A in the future, if other service providers could see B's actions, B may be concerned about their *reputation* and so honour A's trust. The formal treatment of repeated interactions is provided in the Box: Formal Treatment of Relationships and Reputation.

Formal Treatment of Relationships and Reputation For trust to emerge in the absence of a contract, the game in Fig. 2.3 must be more than a once-off. There must be some probability that it continues. To see how the prospect of an ongoing relationship supports Trust, Honour as an equilibrium outcome, suppose that both parties expect to deal with one another again with probability ρ following any interaction. In this situation, B will consider that ongoing prospect in making their choice between Honour and Abuse in any one period. Let V_B denote the value of that future interaction. Suppose that A plays the strategy that they will trust B only if B has not abused that trust in the past. If B abuses their trust, then A will never trust B again. Thus, if A trusts B, then B receives $v - p + \Delta + \rho V_B = v - p + \Delta$ if they abuse that trust in any game (as $V_B = 0$ in this case) while B receives $v - p + \rho V_B$ if A continues to trust and B continues to honour when the opportunity should arise. In this case, $V_B = v - p + \rho V_B$ or $V_B = \frac{v-p}{1-\rho}$. Thus, if $\frac{v-p}{1-\rho} \geq v - p + \Delta$ or $\rho \geq \frac{\Delta}{v-p+\Delta}$, B will continue to honour A's trust and A will choose to trust B.

We can model trust through reputation in a similar manner. Suppose that B's type is given by Nature—that is, there is a share, μ, of buyers that are 'hard-wired' to never abuse someone's trust. If A only has a single interaction (i.e., the game is played once), then

A's expected payoff from trusting B is $\mu(p-c)+(1-\mu)(p-\Delta-c)$. Thus, A will choose to trust all buyers if this is positive; i.e., that $\mu \geq -\frac{p-\Delta-c}{\Delta}$ (which may hold as Δ could be as large as p). Now, suppose that B doesn't interact with the same seller (that is, A) again but may interact with another seller in the future with probability ρ. Suppose that others can observe how B treats others, and sellers will refuse to deal with B if they see B abuse another seller's trust. (This might be akin to a bad credit rating.) In this case, if B abuses a seller's trust, their payoff is $v - p + \Delta$, while if they do not abuse their trust, their payoff is $v - p + \rho V_B$. This has the same structure as ρ derived above for the relationship case. Thus, reputation becomes valuable as a means of facilitating trust so long as ρ, the probability of a future interaction for B, is sufficiently high.

The challenge for relationships and reputation as trust amplifiers is that they rely on there being more activity outside of the transaction with others who can observe what happened within the transaction.

Enforceable Contracts

Where trust may be absent or insufficient to be relied upon, societies have created institutions to enforce contracts. Instead of just coming to an agreement over price, A and B could enter into a formal contract. This typically involves A and B putting their agreement 'in writing.' Thus, at time, t_0, A and B sign a document that states that B will pay A, p, if A supplies a service to B. The idea is that should B abuse A's trust, A will be able to take the document to a Court who, in turn, will compel B to pay A. A then need not rely on trusting B to honour the agreement as a third party; a judge will have the power to ensure A is paid by B.

Unfortunately, this process is not without cost. There are court costs, although, in some countries, these may be reimbursed to A should they prove successful in their case against B. More critically, can A rely on the court holding the contract as valid? In particular, when the written contract is digital, there may be concerns that it could be easily altered. While it would likely be impossible for B to produce an altered contract with no payment for A's service, the payment could be reduced because the Court would not be able to verify that the contract terms were perfect. Alternatively, A could require a third party to be present when the contract is signed and to hold a verified copy of the contract. This would involve incurring costs at the time of signing. Taking all of this into account, even when they act to minimise these costs, A may expect their costs of going to Court might be $c_E > 0$, which gives them a probability of success of $\rho \in [0, 1]$. Even with a potentially enforceable contract, if $\rho p - c_E < c$, A will choose not to enter into a transaction with B.

Cheap Verification

As we have already discussed, blockchains have the quality that they are a ledger with a built-in audit trail. If they are designed properly, something we will examine in the coming chapters, that audit trail is immutable in that agents observing entries can be assured that those entries were placed there by signatories at the time recorded.

Consider what would happen if, instead of either not producing a formal contract or relying on Court enforcement of any contract claimed to be the record of the agreement by both parties, A and B digitally signed their agreement at t_0 and recorded it on a blockchain—as a hash of the digital document, their digital signatures and a time-stamp. In this situation, suppose that B attempted to reduce the price of the contract at some time $t > t_0$. Because the document was recorded on the blockchain, it could be instantly produced to be verified by any third party. Note that the contract still requires Contract Law to be enforceable, but the costs of verifying its terms and existence would now be minuscule. A court could easily enforce the contract, dramatically reducing c_E or increasing ρ, or perhaps the contract could be enforced by an automated process—typically referred to as a smart contract. When verification is cheap, the costs of enforcing contracts fall correspondingly.[5]

In this environment, a party at risk of abuse is protected by the low costs of verifying an agreement and, hence, enforcing that agreement. Faced with that, B will not find it profitable to abuse A, especially if there were damage payments associated with contractual non-compliance.

Thus, the value of the blockchain is reducing the costs of verification. This can be of use when it is otherwise costly to enforce formal contracts or if it is difficult to use relationships or reputation to otherwise establish

---

[5] See Catalini and Gans (2020) for a comprehensive discussion of cheap verification.

trust. Critically, the blockchain cannot create contracts or contract terms that don't exist and cannot enforce items that cannot be written down in a contract or perhaps hard-coded as part of a smart contract.[6] However, it can enable contracts that are feasible within the scope of contract law by reducing, perhaps to zero, verification costs.

What makes this possible is the consensus provided by blockchain technologies. Consensus is an agreement regarding what is being confirmed at a particular time to a blockchain. In the case of a contractual document, the agreement is (i) between the contract parties who sign the document and the transaction that is broadcast to the nodes, (ii) the node that confirms the block containing the transaction and (iii) the other nodes who, depending on the protocol, explicitly or implicitly accept the block. This consensus is what makes the original document's existence and time-stamp verifiable to others in the future. As we will see in future chapters, reaching and maintaining consensus is far from an automatic process but when it is reached the effect is to have a transaction whose nature is verifiable for as long as the blockchain is available.

In summary, while trust can arise and be maintained in the absence of enforceable contracts, this only arises when the probability that the agent who might abuse another's trust expects to interact with the other agent or someone who has observed their interaction in the future. If this probability is low, then either no transaction occurs, or it occurs using enforceable contracts that may limit the types of transactions that are worthwhile.

What the blockchain potentially does, by generating consensus over an audit trail, is reduce the costs, in particular, the verification costs, associated with enforceable contracts. This may allow some transactions that would not otherwise take place to take place (that is, there is an expansion on the extensive margin), or for some transactions, it may cause those transactions to rely on contracts rather than relational or reputational trust alone (that is, there is an expansion on the intensive margin).

---

[6] There is potential for blockchain smart contracts to allow various mechanisms to be implemented that themselves could expand the space of feasible contracts. See Gans (2022) and Holden and Malani (2021) for analyses of this potential.

Cryptocurrencies: Verification-Enabled Payments

This discussion of the drivers of the value of blockchain consensus has not yet considered the main immediate novel use-case of blockchains since Nakamoto (2008), the provision of cryptocurrencies. Cryptocurrencies are tokens (aka digital assets) that are made rival and so can only be owned by one agent at a time (or more precisely, one wallet at a time as the wallet may be owned by more than one agent). What is important to realise is that it is verification that makes cryptocurrencies and the ability to use them as payments possible.

In our earlier example, whereby Ariel sends Bailey 3 tokens as a payment (or for any other reason), how can Bailey be sure that Ariel owned those three tokens in the first place? The answer is that when Ariel came to be in possession of w_A tokens at time t_0, this was the result of someone(s) else creating and confirming a transaction to a block at $t < t_0$ that was confirmed by consensus at that time. This meant that when Ariel came, at $t > t_0$, to place 3 of those tokens in a transaction that moved them to Bailey's wallet, that transaction would only be valid if Ariel digitally signed the transaction (using their private cryptographic key) and Ariel's wallet contained at least three tokens at $t > t_0$. In other words, the transaction required that it could be assured that Ariel had at least three tokens which is something that had been verified via past consensus.

Verification is at the core of all payments—not just cryptocurrencies. Payments in cash are the easiest, as possession of the coins or notes can be physically verified at the time they are being used for payment. Digital debits from a bank account are verified when merchants use a machine to query the consumer's bank that there are sufficient funds in the account to cover the payment whereupon payment immediately occurs. In this situation, the verification is supplied by the bank (or the bank's computer). Cheque payments are interesting because they do not involve immediate verification. While it may be possible for a merchant to call a bank and verify whether the checking account has sufficient funds, in many cases (for cheques with small amounts), this did not happen. Here, the merchant would trust the consumer and could potentially miss a payment if there were insufficient funds to cover the payment (i.e., the cheque would bounce).

Cryptocurrencies are, therefore, just a different mechanism for verification that sufficient funds are available for a payment to be made. However, that verification comes from potentially many different agents rather than just the paying party or their bank. Moreover, the conditions for that verification are built into the blockchain protocol itself. Nonetheless, a concern emerges if Ariel, in our example, were able to control sufficient numbers of nodes so that they could restore the funds to their wallet by eliminating evidence of the transaction that transferred those funds to Bailey. If Ariel could complete their full contract with Bailey (e.g., by receiving goods, services or other assets) and then eliminate that transaction, Ariel could spend those tokens again; this is the so-called 'double-spending problem.' This would leave Bailey having to engage with Court-based enforcement of the contract (if one existed) to be paid. Otherwise, Bailey would need confirmation that the payment transaction from Ariel was immutable. As we will discuss in a later chapter, different blockchain protocol designs give rise to different levels of confidence for someone like Bailey that their receipt of funds has been confirmed; that is, that there is finality in consensus.

2.3   CONSENSUS AND TRUST

Records written to the blockchain have value in providing verifiability of their contents so long as they are believed to be immutable. Specifically, when a dispute arises at time t_1, the parties should be able to examine the relevant record placed on the blockchain at an earlier time $t_0(< t_1)$ to verify claims that may arise in the dispute. Importantly, because an earlier record requires the digital signatures of the parties, even if a record were removed at some $t \in (t_0, t_1)$, it cannot be replaced by a tampered record with different contents. In other words, unlike databases or ledgers, the append-only structure means that any tampering with the record can only result in the removal and not the replacement of a record where both parties must be signatories.

Immutability, therefore, refers to records being preserved and not removed. When there is only a single copy of those records, there is a risk a record might be 'lost.' By distributing the blockchain amongst many nodes, that risk is mitigated. Not only does it become unlikely a record is lost but also makes it less likely that intentional removal occurred.

This latter benefit is quite subtle and worth spelling out in detail. Suppose that someone had an incentive to remove a record placed at t_0 at some time before t_1. One approach would be to remove the record at t_1. This would require proposing a new block at $t \in (t_0, t_1)$, B_t, that was appended to a chain that removed the record at t_0. In this case, the block hashes would have to be rewritten between t_0 to t. With the blockchain up to B_{t-1} distributed, all blocks between t_0 and t would be altered. The probability that such rewriting would go unnoticed by one or more nodes is very low. However, that probability is decreasing in t; that is, the closer the time of tampering is to the time when a record was made, the easier it is to rewrite the record between t_0 and t. The flip side of this, however, is that the contractual obligations being performed by one or more parties relying on the existence of the record at t_0 are less likely to have been completed if t is close to t_0. This will be relevant if parties check for the record's existence at a time after t_0 where they are actually relying on the record.

This may seem a little cumbersome in terms of an explanation so let's consider it for the classic case of a *double-spend attack*. This type of attack is one that we will come back to because it represents a canonical issue for the verification of cryptocurrency ownership. But ultimately, it is an attack that involves the attempted removal of a past record. For example, suppose Ariel has one token and uses it to purchase a service from Bailey. Ariel sends a token to Bailey ahead of obtaining the service. After Ariel is supplied the service, the transaction record of the token sent by Ariel to Bailey is erased. Ariel then has possession of the token again and uses it to purchase something from Casey. By erasing the transaction between Ariel and Bailey, Ariel can spend the token again as the state of the ledger now shows Ariel in possession of a token.

This possibility was addressed by Satoshi Nakamoto (2008). To ensure that Bailey really possessed the token before the service was supplied to Ariel, Nakamoto argued that Bailey should wait for an additional number of blocks to be confirmed before providing the service. In other words, the protection against a potential double-spend (assuming that removal was possible) was to wait a certain number of periods after t_0. In that way, Bailey was protected by the continued blockchain consensus regarding token possession.

The more independent nodes come to a consensus regarding the blockchain and are doing it repeatedly over time, the lower the probability that a record might be removed. Recall that ρ was the probability that a contractual obligation can be enforced. Taking into account the impact of a lost record on that probability, we can posit that probability is now dependent on n, the number of independent nodes. In particular, $\rho(n)$ is increasing in n. Moreover, it is also the case that the addition of one node increases that probability by less if there are a greater number of nodes; that is, $\rho(n+1) - \rho(n) \leq \rho(n) - \rho(n-1)$ for all $n > 0$.[7]

By contrast, in the absence of a blockchain, the relevant record may be held by one agent other than the parties. Thus, the probability that a contract obligation is successfully enforced is $\rho(1)$ for non-blockchain trust. The lower the $\rho(n) - \rho(1)$, the smaller the gain in value the parties will receive from using the blockchain. The case of interest is where the blockchain enables contracts that would otherwise fail because of poor enforceability, i.e., where $\rho(1)$ is so low that at least one party chooses not to enter into the contractual arrangement. In terms of our earlier example, with the blockchain, the critical condition for B, who is concerned about payment from A, is that $\rho(n)p - c_E \geq c$.

Thus far, we have stated $\rho(n)$ as a primitive of the model. In actuality, the relationship between the number of nodes coming to a consensus and the enforceability of the arrangement will depend on the precise workings of the blockchain. For instance, if the contractual risk between A and B is that A might double-spend (which is the equivalent of not honouring B's trust as specified earlier), then $\rho(n) = 1$ if it prevents that double-spending from occurring. This is something we will explore in detail in subsequent chapters.

Key insights from Chapter 2

- Blockchains are a special class of distributed ledgers where many nodes both hold the blockchain data and are responsible for updating it but do so by appending adjustments to existing entries.
- Blockchains enable the time-stamping of blockchain updates which could include when new information is recorded. This creates opportunities to time-stamp digital documents and assets in a reliable way.
- The value of blockchain consensus is that any node can be queried with respect to database entries and records. This allows for lower costs and automated verification of contractual obligations. This also allows for verified holdings of cryptocurrencies.

---

[7] Strictly speaking, ρ may depend on n and $t - t_0$ where t is the time after t_0 that the contract obligation is performed (e.g., supplying a service). Using Nakamoto's logic, ρ is increasing in $t - t_0$. Thus, in choosing t, the contracting party is trading off enforceability with the delay in receiving the final contractual payoff.

> • With cheaper verification, contracts can be written and enforced in settings where otherwise trust mechanisms such as relationships and reputations would be required.

References

Catalini, C., & Gans, J. S. (2020). Some simple economics of the blockchain. *Communications of the ACM, 63*(7), 80–90.

Gans, J. S. (2022). The fine print in smart contracts. In M. Corrales Compagnucci, M. Fenwick, & S. Wrbka (Eds.), *Smart contracts technological, business and legal perspectives* (Chapter 2). Hart Publishing.

Haber, S., & Stornetta, W. S. (1990). How to time-stamp a digital document. In *Conference on the theory and application of cryptography* (pp. 437–455). Springer.

Holden, R., & Malani, A. (2021). *Can blockchain solve the hold-up problem in contracts?* Cambridge University Press.

Nakamoto, S. (2008). Bitcoin: A peer-to-peer electronic cash system. *Decentralized Business Review*, 21260.

CHAPTER 3

Security Versus Speed

Abstract This chapter explores the trade-offs between security and speed in blockchain consensus protocols. It focuses on two primary consensus approaches: Byzantine Fault Tolerance (BFT) and the Longest Chain Rule (LCR). BFT involves multiple communication rounds between nodes for block confirmation, offering robustness against faults at the expense of speed. LCR relies on a block proposer to communicate confirmed blocks, potentially creating forks but allowing faster block confirmations. The chapter emphasises the importance of balancing security and speed in consensus protocols to achieve a functional and trustworthy distributed ledger system.

Keywords Blockchain · Consensus protocols · Byzantine Fault Tolerance · Longest Chain Rule · Security vs. speed

The cornerstone of blockchains are their consensus protocols. Recall that a blockchain is an example of a distributed ledger which means that distinct and independently operating nodes hold and update copies of the ledger. The ideal use case for such a distributed ledger is that anyone can potentially query an entry on the ledger held at one node and be confident that the results of that query are the same from running the same query at any other node. In other words, despite being distributed, from a

© The Author(s), under exclusive license to Springer Nature Switzerland AG 2023
J. Gans, *The Economics of Blockchain Consensus*,
https://doi.org/10.1007/978-3-031-33083-4_3

user perspective, each agent can treat a copy of the ledger as if it were the one 'official' version that would arise in a centralised (or non-distributed) system.

With a blockchain, as it is an append-only ledger, when a block of transactions, content or as we will generally refer to them, 'messages,' are added to the chain, to ensure that user queries across nodes are identical, all nodes must confirm the same block at the same time. When the same block is confirmed by all nodes, then consensus has been achieved. That consensus is with respect to both the content of the block in terms of messages but also the time recorded for each unit of content. The primary (but not the only) challenge to achieving consensus is that communication of the underlying messages and confirmed block does not happen instantaneously. It takes time. Thus, one objective of a consensus protocol is to minimise the time taken to reach a consensus regarding a new block.

One way consensus can eventually be reached is for each node to add messages to a block as they receive them. So long as each node receives messages in the same order, consensus will eventually be reached without nodes needing to communicate the blocks between them. Order is important because if two proposed messages create a conflict, the one received first will, by convention, become the entry, and the other must be discarded. However, this outcome is unlikely to be achieved in a network as messages are received by some nodes who broadcast them to other nodes. Frictions in that process mean that it is unlikely any two nodes, let alone all of them, would receive messages in the same order.

Another way to reach a consensus is to have a block proposed by one node and for it to be confirmed if all nodes unanimously agree that the block should be confirmed. Once again, communication frictions make this more theoretical than practical. In particular, some nodes may be *faulty*. A faulty node may be offline, or it could also be operated by a node with preferences that may be dishonest or malicious (that is, not aligned with the interests of the network).

While users would want a blockchain network to be secure (i.e., not have anything bad happen) and to be what computer scientists call 'live' (i.e., have something good happen), these two goals may be in conflict with one another. Trivially, a secure network could be achieved by not confirming any blocks, but this would conflict with being live, which requires blocks to be confirmed. Frictions in communication make the trade-off between security and liveness or timeliness even sharper.

Computer scientists have spent decades designing protocols that can achieve consensus in distributed ledgers. Typically, they involve a selection process that selects one node as a leader to propose a block, followed by a communication process amongst other nodes that, in turn, are guided by rules to arrive at an eventual consensus. Here we will explore the two main approaches or protocols for consensus: Byzantine Fault Tolerance and the Longest Chain Rule. The former was developed for general distributed networks, while the latter was specifically proposed by Nakamoto (2008) for blockchains. The goal here will be to ground each protocol in economic terms so as to properly identify trade-offs between them in terms of security and timeliness.

The basic model setup involves individual agents, $i \in I$, who send transactions or messages, $m_i \in M$ that are assembled by nodes, $j \in \{1, ...j, ...n\}$, into a block at time t, B_t. It will be assumed here that there is a randomisation process that selects one node in each period to be the proposer or leader who proposes a block, B_t, to be confirmed or accepted by other nodes.[1] Critical to reaching consensus is the communication process amongst nodes and with the leader. In many respects, this communication process resembles a 'cheap talk' game studied by economists.

3.1   Byzantine Fault Tolerance

Byzantine Fault Tolerance (or BFT) has its origins in an analogy used in computer science that relates the problem of achieving consensus to the challenge of coordinating an attack. A typical statement of the Byzantine Generals' problem is as follows:

Two divisions of an army are camped on two hilltops overlooking a common valley. In the valley awaits the enemy. It is clear that if both divisions attack the enemy simultaneously they will win the battle, whereas if only one division attacks it will be defeated. The generals do not initially have plans for launching an attack on the enemy, and the commanding general of the first division wishes to coordinate a simultaneous attack (at some time the next day). Neither general will decide to attack unless he is

---

[1] The existence of a randomisation process is often assumed in computer science, but it is generally accepted that no computer-generated randomisation is truly random. Here we will follow the normal computer science analytical approach.

sure that the other will attack with him. The generals can only communicate by means of a messenger. Normally, it takes the messenger one hour to get from one encampment to the other. However, it is possible that he will get lost in the dark or, worse yet, be captured by the enemy. Fortunately, on this particular night, everything goes smoothly. How long will it take them to coordinate an attack? (Halpern & Moses, 1990, pp. 555–556)

We can label the two generals A and B. If A decides they want to potentially attack, they send a message to B of 'attack.' The problem is that A cannot be certain that the message has reached B. B understands A's need for acknowledgement of the message's receipt and so sends a return message to confirm receipt. Now the problem flips to B who knows A will not attack unless the message receipt is delivered, but B cannot be confident that is the case. A thus acknowledges B's receipt and so on. Those familiar with game theory might notice that this situation is related to the usual assumption of common knowledge (i.e., that it is assumed that A knows that B knows and so on). Indeed, Rubinstein (1989) used a similar context to surface a challenge to the assumption of common knowledge in game theory.

As the problem here is stated, there appears to be no way the generals, despite their common interests, could coordinate an attack. The economics approach to these issues of common knowledge has been to assume common knowledge. By contrast, the computer science approach has been to consider coordination as being conducted by programmed algorithms that do not engage in higher-level thinking. Thus, they specify protocols in which such programmed nodes might reach Byzantine agreement and then consider factors that might disrupt those processes.

Idealised Environment

The possibility of Byzantine agreement starts with an idealised environment where all nodes are operating correctly (i.e., they are not faulty) and the use of the (extensive form) game theory assumption that the order of moves by nodes is specified and known; an assumption called *synchronicity*. Finally, all messages from agents are signed by them using digital signatures, so it is not possible for any agent to represent themselves as another agent.

In this situation, Dolev and Strong (1983) proposed a protocol that is shown to reach an agreement under various types of faults that might occur. Suppose that there are n agents coordinating an attack. The payoff to each agent if all agents coordinate (with an attack or not) is greater than the payoff if agents fail to coordinate.

1. In round 1, one agent, the sender, is chosen at random and sends a (signed) message, $m \in \{attack, not\}$ to all $n - 1$ other agents.
2. In round $t = 2$ to $\min\{f + 1, 2\}$ (where f is an arbitrary number less than n), any agent who has received a message signed by the sender and $t - 1$ other distinct nodes sends that message and the list of signatories to all other nodes.
3. In round $\min\{f + 1, 2\}$, if an agent has sent only one message value to other non-senders, then obey the instruction given. Otherwise, choose not to attack.

Proposition 1 Suppose that $f = 0$, the (Dolev-Strong) protocol leads to an agreement to attack if the sender wants to attack and not to attack otherwise.

Proof With $f = 0$, the protocol lasts two rounds. If the sender sends a message to attack, all non-senders broadcast that message with their signatures appended to all other agents. As all agents will follow the sender's instructions, they avoid the payoff associated with a failure to coordinate. The same logic holds if the sender sends a message not to attack.

Thus, we have the desired situation that all agents reach an agreement.[2] Moreover, just a single round of communication is required.

What, then, is the potential relevance of f that causes the protocol to have more than one communication round? The answer is that some, specifically at most f, of the agents may be faulty. Faults may be of different types. One natural type of fault is where agents *crash*. In this case, an agent may operate normally until some point when it crashes and no longer receives or sends messages. If it were (somehow) known

---

[2] A similar coordination outcome was explored by Farrell and Saloner (1985). In their model, all firms in an industry are choosing whether to switch to a new standard from an old one. They make their choices in a specified sequence. So long as each firm prefers to switch if firms previous to them have switched, then the first firm will choose to switch as it anticipates (correctly) that, by doing so, all other firms will switch. Thus, coordination failure is not possible. They then show that this same outcome results even if firms can choose their place in the 'decision order.'

that at most f agents could crash and coordination by at least $n - f$ agents was sufficient not to be considered a failure, then no adjustment to the protocol is required to avoid a coordination failure. That said, it still requires multiple rounds to confirm this. For instance, suppose that one of three agents in a network may be faulty. Ariel may send messages to Bailey and Casey to attack. Bailey receives the message but does not know whether Casey has because Ariel may have crashed before sending that message. The second round then allows Bailey to confirm what message Casey received. Only in this way will Bailey know that the two non-faulty agents agree to attack.

Malicious Nodes

More concerning than crash faults is the possibility that some agents may not act according to the protocol. The Byzantine Generals' problem becomes more challenging when it is possible that one of the generals might be a traitor. A traitor does not share the preferences of the other honest generals and would prefer a situation where there is a coordination failure with an insufficient number of agents attacking, resulting in defeat. The question is whether, if honest agents follow the protocol but perhaps only $n - f$ are honest, can those agents engineer a coordination failure? For distributed systems, this would mean that the network failed to reach a consensus.[3]

While for the Byzantine Generals' game, it is easy to provide a narrative as to why an agent might have traitorous preferences, computer scientists tend to explore the security of protocols by imagining what a purely malicious agent or agents might be *able* to achieve. Thus, it is important to keep in mind when exploring security issues in this way that, typically, an economist would look for some theory or evidence that an agent with such malicious *intent* might actually exist.

To build intuition, suppose that it is common knowledge that one of the agents is malicious (i.e., $f = 1$). Then, if that agent is not selected as a sender, this would not change the outcome from the protocol as the only message would originate from an honest node. But what if the

---

[3] These malicious agents are sometimes referred to as 'Byzantine agents,' which is strange since it was the Byzantine generals who were trying to coordinate an attack on their enemy. Thus, they should be more correctly referred to as 'non-Byzantine agents,' but somehow computer scientists have let the terminology evolve in this strange way.

malicious agent is selected? Recall their incentive is to create a coordination failure which means (a) having a set of honest non-senders receive a signed message from them to attack and the remainder receiving a message not to attack; and (b) that those non-senders are not confused by the end of the protocol; that is, they only receive evidence of one type of signed message from the malicious sender and so do not have conclusive evidence that the sender is malicious.

It can easily be seen that one malicious agent cannot engender a coordination failure. To see this most simply, suppose that $\frac{n-1}{2}$ agents are sent a message to attack in round 1, and the remainder are sent a message not to attack. Then each will send that message to all other agents in round 2 (which is the $(f+1)$st round). All honest agents would, therefore, receive conflicting messages and not attack. They have reached an agreement, so there is no coordination failure.

What if, instead, $f = 2$? As we have seen, to test security, computer scientists are prone to imagine worst-case scenarios. Thus, even if no honest agent can tell whether other agents are honest or not at the outset, it is assumed that malicious agents know who is malicious and can, therefore, collude in their messages to unsuspecting honest nodes. Specifically, now, when one of the malicious agents is a sender, they can send all honest agents a message to attack and a message not to attack to the malicious non-sender. In the next round, the malicious non-sender sends their receipt to a subset of the honest non-senders. Those non-senders now have conflicting messages and know the initial sender is malicious. The other non-senders do not receive a message from the malicious non-sender and so only have signed messages from $n - 2$ nodes to attack.

Herein is where what is known as the Dolev-Strong protocol thwarts the attack. When it is known there could be two malicious agents; there is an additional communication round. In that round, the honest non-senders, who have just received a message not to attack, now broadcast that double-signed message to others. Thus, all honest non-senders end up with conflicting messages and so do not attack. In effect, the additional round means that the malicious sender has now been exposed to all.

The following proposition demonstrates that coordination failure will not arise for any number of malicious agents, f.

Proposition 2 The (Dolev-Strong) protocol leads to an agreement to attack if an honest sender wants to attack and no agreement to attack otherwise.

Proof If the sender is not malicious, they have a single signed message to all agents, and thus, only that message can be propagated and followed. Moreover, there is no possibility of confusion.

Suppose that the sender is malicious. Before round $f + 1$, whenever an honest agent receives a new message that is signed by the sender and the requisite number of distinct agents, they will be able to send that to all other honest non-senders and so thwart a coordination failure. In round $f + 1$, if an honest non-sender receives a new appropriately signed message, they cannot send it to others. However, because that message is signed by $f + 1$ others, it must be signed by an honest agent who can be inferred sent that message to others previously, perhaps in the same round $f + 1$. Thus, all honest non-senders reach an agreement, and coordination failure is not possible.

This demonstrates that, regardless of $f < n$, this protocol leads to consensus in that after $f + 1$ rounds, all non-faulty agents will attack or none will.

It is useful to reflect on why the Dolev-Strong protocol prevents a collusive malicious attack. Remember that each agent will only accept (and then broadcast and potentially follow) a message if it has been signed by the sender and then signed by another distinct agent in every communication round. If an agent receives a message satisfying that property, they communicate it to others if there are rounds remaining. If that is the only message they have received up to that point, their presumption is that they want others to see that message so they can check if there are conflicting messages in the network satisfying those conditions. If there are, then the agent knows the sender is malicious and will choose to ignore those messages. If there are not, then the agent knows there are no conflicting messages being broadcast so long as they are certain there is one honest agent amongst the message with the most signatures. Having $f + 1$ communication rounds provides that certainty.

Thus, it is a combination of requiring signatures and the strength of that requirement by the end of the protocol that protects the network from coordination failure. The agents can simply be programmed how to react when presented with signed messages as evidence. But there is also a sense in which, if the agents were not programs but strategic agents, their chosen behaviour would be the same. A strategic agent would form a belief that any other given agent was malicious or not. At the outset, their prior would be that an agent is malicious with probability $\frac{f}{n-1}$. This prior is not updated unless (a) an agent receives two conflicting signed messages, in which case the sender is malicious with probability 1 or (b) an agent receives a signed message that they know has been signed by a non-malicious agent who themselves has formed a probability that the sender is not malicious with probability 1. That second step can only occur in the final communication round. Computer scientists refer to this situation as an agent being 'convinced' regarding a message. From an economist's standpoint, being 'convinced' involves having a posterior probability of 1 that the sender is malicious or not. Thus, the computer scientists' use of the term convinced is weaker and is merely a trigger to broadcast a signed message to other agents.[4]

In summary, the ability to sign messages means that messages are not quite cheap talk because sending conflicting messages to different agents runs the risk of being identified as being malicious. The protocol forces that cost to be realised by having a sufficient number of rounds of communication. It is here where we can see that timeliness is sacrificed in other to make the network secure. The more malicious nodes there are believed to be, the greater the number of communication rounds and hence the time it takes to reach an agreement.

No Digital Signatures

In many blockchain protocols, digital signatures (and associated public-key infrastructure) are not deployed for the identification of specific nodes. This may be for ideological reasons, in terms of allowing anyone to participate as a node, or for computational reasons, as checking identity for each round of communication involves additional costs. However,

---

[4] A game theorist would typically assign some payoffs to the good outcome, coordination, and the bad outcome, failed coordination; say, V and L respectively. Then as honest agents receive a message of length t, they would have an updated probability that the node they hear from is malicious or not. For instance, if the message is of length 2, then the probability that the sender is malicious is $\frac{f}{n-1}\frac{f-1}{n-2} + (1 - \frac{f}{n-1})\frac{f}{n-2} = \frac{f}{n-1}$. Thus, the prior equals the posterior probability until $t = f + 1$. Halaburda et al. (2021) provide a comprehensive game theoretic analysis of the single round mechanism with explicit preferences but also agents who are ambiguity averse, which is one way of capturing the type of 'worse-case scenario' computer scientists focus on.

as we have seen, not having digital signatures makes it more difficult to identify malicious nodes because it prevents honest nodes from accepting and then spreading the messages of others.

To see how a lack of digital signatures creates challenges, suppose that Ariel is a leader and proposes an action to Bailey and Casey. If all agents are honest, then upon hearing the message from Ariel, both Bailey and Casey relay it to each other, and consensus is achieved. But what if (a) Bailey is not honest and (b) Bailey does not want to do the action proposed by Ariel? In this case, upon receiving Ariel's message, Casey relays that message to Bailey, but Bailey wants to relay an alternative message, they claim to be from Ariel, to Casey. With digital signatures, Bailey cannot do that as Ariel's signature cannot be changed, and hence, it cannot be claimed by any node that Ariel's proposal was any different from their intended message.

Without digital signatures, Bailey can get away with the subterfuge. Casey then cannot distinguish between two possibilities: (i) that Bailey is dishonest and the message from Ariel is true or (ii) Ariel is dishonest and is trying to prevent consensus by sending conflicting messages. Thus, a consensus is not reached, and the agents do not know what actions to take to avoid coordination failure. Put simply, any honest node can tell if there is a dishonest node amongst the other two but not which node is dishonest.

In computer science, it was demonstrated that this intuition generalised to larger networks and that there were protocols that did not require digital signatures that would reach consensus so long as $f < \frac{n}{3}$ and that consensus could not be guaranteed if $f \geq \frac{n}{3}$ (see Fischer et al., 1986; Pease et al., 1980). Specifically, the protocol involved Byzantine Broadcast, along the lines of the protocols already examined here, where a sender is first selected and moves first, and had the properties of terminating (i.e., eventually halting with some output from each node), agreement (with all honest nodes producing the same output) and validity (if the sender is honest, then the output of honest nodes is that of the sender).

Consider this proposition stated in a similar manner to the propositions above.

Proposition 3 Suppose there is a protocol where a sender is selected at random and non-sending agents confirm an output once they have received $n - f - 1$ messages from non-senders. Then so long as $f < \frac{n}{3}$, there is an agreement to attack if an honest sender wants to attack and not to attack otherwise.

Rather than provide a formal proof that involves the construction of simulated nodes as discussed by Fischer et al. (1986), here the particular disruptive strategies of malicious agents are considered within the Byzantine Generals' game.

First, suppose that a malicious agent is selected as a sender. Recall that agent wants fewer than the $n - f$ honest agents to commit to attacking. To achieve this, the malicious sender partitions the honest nodes into two distinct sets. The idea is to send different messages to each set; one receives an attack message, and the other receives a message not to attack. The honest non-senders in each set will then send that message to other nodes. Given that there was no setup of the network with digital signatures to identify participating nodes, the issue arises of whom to send the messages to. This information resides with the sender, who not only sends their message but also the IP addresses of other recipients of that message. Thus, each non-sender receives $n - 1$ IP addresses of others in the network. Of these (about) $\frac{1}{2}(n - f)$ are honest agents in each partition. The remaining f agents are part of the malicious coalition with their IP addresses listed in each.[5] Thus, when the malicious sender sends different instructions to each partition, the malicious agents in each partition themselves send different confirmations to different partitions. Thus, any honest agent in a partition obtains $\frac{1}{2}(n - f) + f$ confirmations of the message they receive from the sender. As this is less than n, what would honest agents make of this? An honest agent knows that $n - f$ agents are honest and will definitely relay the sender's message. Thus, they will realise something is amiss (or strictly speaking, their protocol will not halt) if the total number of agents they receive notifications from is less than $n - f$ or $\frac{1}{2}(n - f) + f < n - f \implies f < \frac{n}{3}$. However, if this condition does not hold, then the honest agents are potentially fooled, with a subset of them choosing to attack with a resulting defeat.

Second, what if an honest agent is selected as the sender? In that case, malicious agents can send conflicting messages relayed to different sets of honest non-senders. If the honest agent's message was to attack, this means that $\frac{n-f}{2}$ honest agents will receive $\frac{n-f}{2} + f$ messages to attack

---

[5] Why are these f agents and not $f - 1$ agents? The reason is that the malicious node who is a sender, can also simulate being non-sending nodes.

while the others will receive conflicting messages and choose not to attack. Those who are part of the 'attacking' partition will know something is amiss if their total messages are less than $n - f$ or $f < \frac{n}{3}$. Thus, if there is a share of malicious agents, they can, as non-senders, fool only a subset of honest agents to attack, resulting in coordination failure.

This proposition demonstrates that synchronous networks can tolerate faults up to 33 per cent of nodes. As we will see, this has inspired the construction of blockchain protocols based on BFT using Proof of Stake as a basis for the selection of validating nodes. However, the efficacy of these protocols for blockchain consensus critically relies on the ability to use time as a regulator in the protocols. Without this, achieving consensus becomes much more challenging.

Asynchronous Networks

What if nodes in a network do not have access to a common global clock? Having a clock means that the protocol can be defined in terms of time steps that allow rounds to have a clear structure. Thus, any message sent at time t is assumed to arrive to any recipient by time $t + 1$. This allowed us to specify the number of rounds of communication required to reach a consensus and to weed out faulty nodes when messages from them are not received.

The problem is that such guarantees are not reasonable when consensus is being coordinated over the Internet. Even when being generous regarding the length of time for each round, there are many ways in which delays can arise on the Internet from outages, internet service provider problems and denial of service actions that can be taken by malicious actors. The synchronous assumption might be rescued if the length of time periods was extended to handle worst-case scenarios, but this would mean that the network was *always* slow even if worst-case scenarios were rare. For that reason, computer scientists are loathe to assume that time is anything other than asynchronous.

The immediate implication of this is that when a node does not receive a message from another node, it cannot tell if it is just a 'normal' outage or a deliberate choice from a malicious agent not to send a message from that node. This means that, in addition to the ability to send messages as a coalition, malicious agents can now alter message delivery. In computer science, this is called *adversarial message delivery*. The implication of this is that protocols that are fault tolerant need to be robust to malicious agents in terms of message content sent to each recipient but also to the choices of which messages—from honest and malicious nodes alike—are delivered. In economic terms, the protocol is robust to the perhaps (implausible) scenario of a worst-case scenario in terms of malicious agent ability.

With asynchronous networks, it is not possible to construct a protocol that is fault tolerant (Fischer et al., 1985). That is, one malicious agent can cause coordination failure. To see this, recall that honest non-senders do not know what to make of a silent node on the network. Is it an honest sender whose messages have been delayed or a malicious agent who has chosen not to send a message? In the Byzantine Generals' game, all honest agents need to coordinate on an attack or it fails. If there is one malicious agent (i.e., $f = 1$), then $n - 1$ agents need to attack. Thus, if selected, a malicious sender can send 'attack' messages to a partition of honest agents and also control message delivery so that those agents' messages themselves do not reach the honest agents in the other partition. Similarly, control over communications can be used when an honest sender sends an attack message to ensure that there is a partition of agents who never receive that message. What this means is that if they are selected as the sender, a malicious agent could ensure that $n - 2$ agents receive an attack message and confirmations from $n - 1$ agents while one honest node does not receive a message to attack. Even if they are not selected as the sender, a malicious node can control the message delivery to one honest agent and imitate their behaviour with the remaining honest agents to ensure that only $n - 2$ attack. (Alternatively, the malicious agent could impact message delivery to prevent consensus from ever being reached.)

This implies that, in order to have a fault-tolerant consensus protocol, there will have to be some way of overcoming issues associated with the timing of messages. As will be shown in future chapters, the way blockchain protocol designers have overcome this challenge has been to create protocols that allow for partial synchronicity in that while there may be a fixed time period whereby there is asynchronicity, and message order cannot be confirmed, there is a subsequent period that terminates at some, usually unknown, time.

What Determines the Power of Malicious Actors?

The computer science approach to security in coordinating distributed nodes searches for a protocol that achieves a set of outcomes. For a given protocol, the idea is to consider whether the protocol is robust to the presence of faulty nodes. It is not robust if those nodes result in an inability to either reach an agreement (all honest nodes confirm the same information and hence, take the same action), validity (that if all honest nodes have the same information, then that information becomes the consensus) and termination (all honest nodes eventually confirm some information and take the requisite action).

As has been shown, the presence of faulty nodes can undermine the ability of a given protocol to achieve agreement, validity and termination. This has been discussed in the scope of the Byzantine Generals' game, where a coordination failure is costly and where fewer than every honest node decides to attack. Clearly, the metaphor of the Byzantine Generals' game is stretched when the goal for a coordination attack involves only honest agents. But the practical goal of consensus, which involves distributed nodes agreeing on relevant information and allowing others to make coordinated decisions based on that information, does not require a 'knife-edge' payoff structure but a structure that some proportion of honest nodes reaches consensus.

That said, the stress testing of any given protocol in computer science involves imagining that the entire set of faulty nodes themselves wants to and can mount a coordinated intervention on the network. While an economic approach sometimes considers situations where such coordinated interventions are possible (for instance, by considering whether a mechanism or equilibrium is coalition-proof) they would not do so without clear assumptions regarding preferences and technology.

For preferences, the Byzantine Generals' story does give malicious agents a motive—they are traitors or infiltrators whose interests are the opposite of honest agents; that is, they want there to be an uncoordinated attack that fails or at least no attack at all. But as already mentioned, the Byzantine Generals' game is a metaphor for network consensus in general. Thus, if we were examining, say, a blockchain network underpinned by a BFT consensus protocol, it would be important to establish what the payoffs were for malicious agents. This would determine whether a failure to coordinate was a motive for those agents or, instead, their motive might be for honest nodes to reach a consensus leading to decisions that were in the malicious agents' interests. On these issues, the computer science literature has been largely silent except for considerations of double-spending.

For technology, if the preferences of malicious agents were specified, the costs of achieving their interventions would have to be clearly spelled out. The BFT literature does not consider such costs. Indeed, it makes a clear assumption that malicious nodes can coordinate their own actions perfectly and, in some cases, manage the flow of information *between* honest nodes. This allows them to stress test protocols against a worse-case scenario of an all-powerful malicious entity. But, for economists, the default would be that malicious agents would be endowed with the same technology as honest ones. In this case, it would be natural to ask, why can malicious agents coordinate their activities to subvert the consensus process of honest agents who are themselves assumed to have no ability to coordinate at all? Surely, a better starting point would be to consider symmetric access to coordinating technologies for all agents—honest or malicious—and then relax those assumptions. By moving to extremes in terms of the distribution of coordinating technologies, the security requirements are surely inflated and it is difficult to evaluate whether they are worth those costs in terms of, say, more communication and more time spent reaching a consensus within a given protocol.

To see why this matters, let's consider the asynchronous model but without adversarial message delivery. Recall that with adversarial message delivery, having one malicious node was sufficient to cause a coordination failure. Without this, if a malicious agent is selected as the sender, they can still send 'attack' messages to a partition of honest agents but now cannot control the messages they send to each other. This means that it is possible (but not guaranteed) that honest agents can identify the sender as malicious if they receive conflicting messages from other agents. Thus, the malicious agent will want to send out the same 'attack' message to a subset of honest agents and nothing to others. But in this case, there is no possibility of coordination failure.

This means that the computer scientist approach based on worst-case scenarios for adversarial intervention tells us that you cannot guarantee both security and speed. The economist approach asks that the powers of the adversary are spelled out, and then a protocol is analysed for security and speed, taking into account the preferences of all involved. In other words, the computer science approach is agnostic to adversary powers, while the economics approach is not. In practice, however, the two approaches end up being closer because, say, protocols will place timeout constraints on message delivery that themselves weaken ambiguous asynchronous environments.

Summarising the Trade-Off

For BFT protocols, the aim is for the protocol to run, reach a consensus and then end. This should happen in each specified time period (which could be on the order of microseconds) and be possible with nodes programmed to exhibit honest behaviour. As has been shown, the simplest two-stage protocol whereby a leader proposes a value and broadcasts that as a message to other nodes would achieve consensus quickly but would be vulnerable to faults. Some of those faults are because nodes can crash. Some of those faults are not really faults but maliciously programmed nodes designed to prevent consensus from being reached.

The simplest way to prevent faults—malicious or otherwise—from subverting consensus amongst honest nodes is for the protocol to take more time. If nodes can be identified with digital signatures and the communication of messages between all nodes can be ordered and time-stamped, then honest nodes can be programmed to assemble a certain number $(f + 1)$ of distinct messages from a sender and become fault tolerant. Thus, rather than taking two rounds of communication, $f + 2$ rounds are required. If digital signatures cannot be used, honest nodes can still be programmed to assemble a certain number $(n - f)$ of messages, but this is only possible if the number of faulty nodes (f) is less than a third of the total number of nodes. If time cannot be used to count communication rounds, then there are situations where even a single faulty node (i.e., $f = 1$) can cause any protocol to not reach consensus.

Thus, time (i.e., reducing speed) potentially makes consensus protocols more secure, and the trade-off between them depends upon whether time can be used to synchronise node-to-node communications and whether digital signatures can be used to identify nodes and their messages. That said, while these protocols can be secure in the sense that consensus is reached, a malicious sender can potentially cause consensus to be reached on an untrue or dishonest outcome. We will see in a later chapter that using protocols based on majority voting as the basis for generating consensus can potentially mitigate a dishonest outcome but that these still decrease the speed at which consensus is reached. Another possibility is to use a mechanism to ensure the truth is messaged even from malicious nodes, and we will return to discuss that in Chapter 7.

3.2 THE LONGEST CHAIN RULE

BFT-style consensus has a long history of being developed for the coordination of distributed computing for many decades. This class of consensus protocols was not developed specifically for blockchain environments (i.e., distributed append-only ledgers) but could be used in them. However, there is another distinct class of consensus protocols developed specifically for blockchains. This class is named the *longest chain rule* (LCR) for the convention it espouses that nodes coordinate and append new blocks to the longest chain they observe.

The LCR was invented by Nakamoto (2008) alongside the proposal for the first permissionless protocol, Bitcoin. Nakamoto was motivated to find a consensus protocol when neither the quantity (n) nor the identity of nodes could be specified. Their conception was to provide a protocol in which any person with a computer could participate. This ruled out protocols that would keep track of nodes that sent and received messages (by the use of digital signatures), any presumption on the number of nodes that may be faulty and the assumption that messages could be sent without delay (asynchronous time). To cut through this, Nakamoto would have participating nodes accumulate messages that were being broadcast throughout the network, assembling them into blocks and then selecting one to immediately confirm their block as the next consensus block on the network. This selection process relied on Proof of Work (something that will be explored in Chapter 5), but, for the moment, it suffices to imagine that a node is chosen at random.

The underlying protocol proposed by Nakamoto of selecting a node at random, broadcasting their block to others and then confirming it (so long as it appropriately references the hash of the previous block) would give rise to immediate consensus in a frictionless environment, where communication was instantaneous, and all nodes were honest. When frictions such as faulty nodes or delayed messages arose, instead of modifying that protocol to include multiple rounds before a block of messages was confirmed (as we saw for BFT protocols), Nakamoto did not depart from the simplest protocol of random selection and broadcast.

The Mechanics of LCR Coordination

So how does the LCR deal with the problems that arise when there are frictions? If messages can be delayed, then it is possible that different nodes will package different messages into any given block. Of course, so long as messages are eventually propagated globally, they may be packaged into subsequent blocks. More seriously, it is possible that the fact that a node is selected and proposes a block does not reach all nodes. In that case, there is no consensus reached on that block. Instead, some nodes may miss a round and then, if they are selected, propose a block that overlaps in terms of messages that other nodes already have confirmed to a chain. This situation is called a *fork* (see Fig. 3.1).

Forks can arise because of frictions in communication—whether it be messages or updates to blockchain code to remove bugs—and also potentially because of the intentional action of adversarial agents who may be seeking to write an alternative set of messages to the blockchain. Because messages have to be cryptographically signed to be valid, such intentional action usually involves removing a message from the blockchain; i.e., some messages are *censored*. Regardless of how they emerged, Nakamoto proposed that forks could be resolved—that is, all nodes eventually work on a common chain—if nodes, when they observe a fork, append new blocks only to the longest chain. This would make that chain even longer and, in the process, cause other chains to stop updating. Those orphaned chains are effectively cast aside.

Will honest nodes converge on the longest chain when a fork arises? As it turns out, it is not obvious they will. Biais et al. (2019) analyse the game that results amongst nodes should a fork arise. To see this, suppose that,

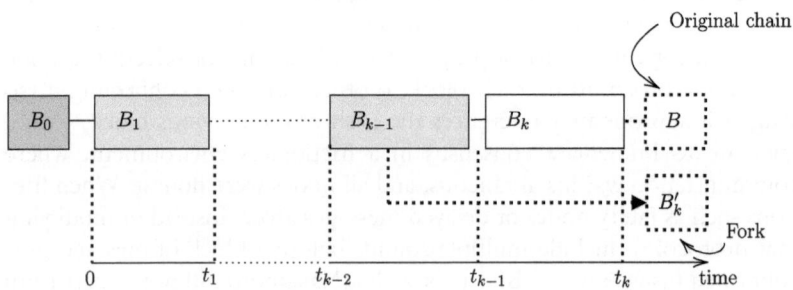

Fig. 3.1 Forked blockchain

at time t, there is a single chain from B_0, the genesis block to B_t. A fork is said to occur at $t + k$ if there exist (B_i, B_k, B'_k) such that $B_k \neq B'_k$ and (B_i, B_k) and (B_i, B'_k) are both chained blocks subsequent to t. Suppose that B_k is part of a chain longer than the chain with B'_k. To complete the statement of the game, suppose that the expected payoff for a node from becoming a leader on a chain also chosen by n_i nodes is $G(n_i)$, which is increasing in n_i. This comes from the reward a chosen leader receives when a block is confirmed on a change that survives in equilibrium.[6]

As it turns out, there are many possible equilibrium outcomes in this game. One is that the nodes all coordinate on the longest chain. If this is done, then, in considering where to mine an additional block, if all other nodes are expected to work on the longest chain, a single node will obtain $G(n)$ by working on that chain and $G(1)$ by working on another chain. As $G(.)$ is increasing, it is easy to see that all nodes would find it worthwhile to follow the LCR.

Interestingly, this is only one type of equilibrium. Suppose that a fork has persisted for some time with a majority of nodes, $K > \frac{n}{2}$ working on one chain and receiving $G(K)$ if they are selected as a leader while a minority work on another chain with block payoffs of $G(n - K)$. Note that $G(K) > G(n - K)$ so that it is better for an unencumbered node to work on the same chain as the majority of nodes. However, as the block has persisted, Biais et al. (2019) show that nodes on the minority chain prefer to stay on that chain as it helps maintain the value of past rewards they have received (i.e., $G(n - K - 1) < G(n - k)$) and they may face lower prospects of receiving future rewards on the alternative chain. This type of equilibrium is, however, Pareto inefficient for nodes as no node earns the maximum possible rewards of $G(n)$.

There have been instances where forks have occurred that have persisted. These happened in relation to differences in opinions amongst nodes regarding how the blockchain might operate. One famous example arose on the Ethereum blockchain when a fund called The DAO was successfully hacked for \$50 million worth of its token, Ether. There was disagreement amongst participating nodes as to whether there should be a roll-back of the chain to restore the tokens to their original owners or whether the consequences of the hack should persist. The end result

---

[6] This reward is typical in blockchains that select a single leader to propose a block. If it is confirmed, the leader may receive transaction fees as well as a block reward in the form of tokens. These will be explored in a subsequent chapter.

was a permanent split into two chains—Ethereum (with the roll-back) and Ethereum Classic (with no change). The latter chain had about three per cent of the participation as the former. Such hard forks have also occurred for Bitcoin over changes to the size of blocks and the way in which Proof of Work operated. Now two chains, Bitcoin Cash and Bitcoin Gold, operate alongside Bitcoin.

Forks represent a useful way in which changes can arise for blockchain protocols. In effect, the longest chain rule acts like an in-build majority voting system. Proposals for changes can circulate, and then some coordinated group announces and orchestrates a fork on a particular date. The success of the fork depends upon whether the majority of nodes support that fork, in which case the LCR ensures that it becomes the dominant chain. For BFT protocols, by contrast, there is no in-built mechanism, and coordination is correspondingly more difficult to achieve if there are significant changes are proposed.

Block Finality

In BFT protocols, when a block is confirmed and added to a chain, it is treated as final; that is, there is no scope within the protocol to censor messages in past blocks. It is, in this sense, that such consensus is seen as secure.[7] To achieve this security for different upper bounds on faulty nodes, the protocol can vary the number of rounds before a consensus is reached. Thus, there the trade-off between security and speed is resolved in the design of the protocol itself.

By contrast, LCR protocols involve a minimal number of rounds to confirm a block. However, confirmed blocks to a particular chain may not be part of other chains if those chains end up being longer. Thus, while the LCR protocol has speed and a consequent sacrifice of security within the protocol, there is an issue for realising the value of blockchain consensus as confirmed blocks may not have *finality*. That is, there is a period in which, if users query a node for blockchain information, they may receive different answers depending on the node queried. True finality occurs when all nodes will return the same information when they are queried.

---

[7] Technically, if all nodes agreed, they could roll back the chain to some earlier block and, in the process, rewrite block history. However, this would require an off-chain deliberation process.

Nakamoto (2008) anticipated this. Unlike a BFT protocol, an adversary may be able to create a fork in the blockchain at some previous point to t, say for block, B'_{t-k}. If the majority of nodes (in terms of their effective voting power) support the fork, then someone who thought a block, B_{t-k} was final could find that is not the case as it is replaced by B'_{t-k}. Thus, no confirmed block—other than the genesis block, B_0—is truly final. That said, the higher is k, the harder it is to fork a blockchain from $t-k$ onwards.

To see this, suppose that there are f adversarial nodes and $n-f$ honest nodes. The probability that the next block at t is chosen as the leader for B_t is p while the probability that it is an adversary is q. The probability, q_k, that an adversary could catch up to the main chain by forking the chain at $t-k$ is:

$$q_k = \left\{ \begin{array}{ll} 1 & \text{if } p \leq q \\ (q/p)^k & \text{if } p > q \end{array} \right\}$$

If the adversary commands the majority of voting power, then their fork will prevail. If not, then there is still a chance it may become the longest chain and the block at $t-k$ would be changed. Thus, for a user relying on a transaction that takes place at $t-k$, the question is at what time that user can become confident that the block at $t-k$ is final?

Nakamoto shows that, at the average expected time to confirm each block, an adversary's progress (which is not observed by honest nodes) is a Poisson distribution with an expected value of $\lambda = k\frac{q}{p}$. Given this, the probability that an adversary will not catch up by t is:

$$P(k) = \sum_{x=0}^{k} \frac{\lambda^x e^{-\lambda}}{x!} \left(1 - (q/p)^{(k-x)} \right)$$

What this shows is that if the user wants to have a higher probability, $P(k)$ that at the time a block is added to a chain, that block is final, then they will want to wait a longer number, k, of periods. For instance, if $q = 0.3$, an agent who wanted $P > 0.999$, would want to wait 24 periods (or for the confirmation of 24 blocks). That said, $P(k)$ is convex in k so that the marginal value of waiting diminishes with time.

What this implies is that if the value, V, an agent expects to receive should there be no adversarial attack, is higher, then the choice of waiting period for finality, k, will be higher.[8] Thus, if someone is purchasing a cup of coffee using a cryptocurrency, a cafe may choose to honour that payment with a cup of coffee immediately whereas if an art house were selling a multi-million dollar work of art, they may wait some period of time before handing the art to a buyer.

This highlights an underappreciated characteristic of LCR protocols compared with their BFT counterparts. For a BFT protocol, finality is achieved when a block is confirmed but the time taken to confirm a block is dictated by the protocol. For an LCR protocol, a block is immediately confirmed but finality takes time. The length of time it takes for finality is *a choice of individual agents*. This stands in contrast to BFT protocols where speed is a 'one size fits all' protocol choice.

3.3   CONCLUSION

If a ledger is not distributed, then so long as the operator is trusted, the ledger is secure, and updates are available immediately. Of course, if the operator cannot be trusted or can be hacked, security fails dramatically. For distributed ledgers, there is a generic trade-off between security and speed. This is because there is a greater chance an individual node is faulty, but also, there is a way of putting in frictions that slow down the speed of updates to the ledger that reduce the impact of faulty nodes.

While computer scientists often characterise different blockchain protocols as being either perfectly secure or maximally speedy, the actual reality of the situation is that the trade-off likely exists more at the margin. This is because the frontier between security and speed (both of which are probabilistic in nature) is concave to the origin. Specifically, it is easier to decrease the likelihood of a disruption to consensus the longer that consensus is expected to be achieved and vice versa.

Thus, from an economic point of view, the optimal protocol will depend on the marginal rate of substitution users have between security and speed. Importantly, this is highly unlikely to favour a protocol that abandons one dimension completely in favour of the other. As we will see in the chapters to follow, this type of trade-off exists in other aspects of blockchain protocol design.

---

[8] That is, if the discount factor per period is δ, the agent chooses k to maximise $P(k)\delta^k V$.

> **Key insights from Chapter 3**
> - There are two broad classes of protocols to reach consensus on blockchains: Byzantine fault tolerance (BFT) and the Longest Chain rule (LCR).
> - BFT involves successive rounds of communication between nodes before a block is confirmed at which point it is regarded as final. The longer the rounds of communication, the more robust the protocol is to faults—whether it be crash faults or malicious actors.
> - The ability to identify individual nodes by their signature and have that signature accompany messages softens the trade-off between security and speed; allowing more of both in BFT protocols.
> - LCR involves the selection of a block proposer who communicates the confirmed block to other nodes. As communication is imperfect, forks can arise. In that case, the LCR approach asks nodes to build only on the chain with the most number of blocks. The possibility of forks makes the blockchain less secure, but this security can be mitigated for users if they delay economic finality for a certain amount of time after a block is added to the blockchain.

References

Biais, B., Bisiere, C., Bouvard, M., & Casamatta, C. (2019). The blockchain folk theorem. *The Review of Financial Studies, 32*(5), 1662–1715.

Dolev, D., & Strong, H. R. (1983). Authenticated algorithms for byzantine agreement. *SIAM Journal on Computing, 12*(4), 656–666.

Farrell, J., & Saloner, G. (1985). Standardization, compatibility, and innovation. *The RAND Journal of Economics*, 70–83.

Fischer, M. J., Lynch, N. A., & Paterson, M. S. (1985). Impossibility of distributed consensus with one faulty process. *Journal of the ACM (JACM), 32*(2), 374–382.

Fischer, M. J., Lynch, N. A., & Merritt, M. (1986). Easy impossibility proofs for distributed consensus problems. *Distributed Computing, 1*(1), 26–39.

Halaburda, H., He, Z., & Li, J. (2021). *An economic model of consensus on distributed ledgers*. National Bureau of Economic Research: Technical report.

Halpern, J. Y., & Moses, Y. (1990). Knowledge and common knowledge in a distributed environment. *Journal of the ACM (JACM), 37*(3), 549–587.

Nakamoto, S. (2008). Bitcoin: A peer-to-peer electronic cash system. *Decentralized Business Review*, 21260.

Pease, M., Shostak, R., & Lamport, L. (1980). Reaching agreement in the presence of faults. *Journal of the ACM (JACM), 27*(2), 228–234.

Rubinstein, A. (1989). The electronic mail game: Strategic behavior under "almost common knowledge". *The American Economic Review*, 385–391.

CHAPTER 4

Permissioned Versus Permissionless

Abstract This chapter examines the trade-offs between permissioned and permissionless blockchain networks. Permissionless networks, like Bitcoin, allow any node to join without prior authorisation but face greater security challenges. To maintain security and deter Sybil attacks, permissionless networks impose costs on nodes and offer rewards, such as block rewards and transaction fees. Comparing permissioned and permissionless systems resembles the trade-offs between monopoly and competition. Permissionless networks encourage diverse participation and potentially higher efficiency but may incur higher costs. The chapter highlights the importance of balancing security, participation, and efficiency in blockchain networks.

Keywords Permissioned · Permissionless · Blockchain · Security · Consensus protocols

Consensus protocols, as have been shown thus far, had their origins in mechanisms to achieve coordination in distributed computing. In those applications, the number of nodes in a network was known—having been designed by a single entity—and the main concern was what happened if some of those nodes were faulty or were under the control of agents with malicious intent. This focused researchers on the notion that the speed at

J. Gans, *The Economics of Blockchain Consensus*,
https://doi.org/10.1007/978-3-031-33083-4_4

which those nodes reached consensus might have to be reduced in order to ensure network security.

Satoshi Nakamoto's vision for Bitcoin shattered those design parameters by positing a distributed network where not only was the identity of nodes unknown, but their number could also not be pre-specified. The underlying idea was that no agent would need permission to join the network and engage in the validation of blocks. The Bitcoin network was proposed to have an open door. Such a permissionless environment posed uniquely difficult problems for achieving network consensus. Thus, it is natural to ask: what are the benefits users would receive from having a permissionless network that would justify the likely greater security issues that would arise relative to permissioned counterparts?

This chapter explores the trade-offs involved in choosing between a permissioned and permissionless network. The first thing to note is that the traditional BFT protocols considered in the previous chapter, where security is established by creating a bound on the number of faulty nodes, cannot be translated easily to the permissionless environment where there may be any number of faulty nodes. The LCR-based protocols, on the other hand, do not rely on such bounds, which is precisely why Nakamoto invented it. But, as will be shown, it is also possible to modify the BFT protocols to operate in a permissionless setting.

In each case, protocols for permissionless networks involve randomly selecting one node to be a leader who proposes a block to be confirmed to the chain. The selection process is, however, constrained in that it should reinforce the benefits (perceived or otherwise) of having a permissionless network. As will be shown, this limits the type of selection rules to determine the leader.

Following leader selection, there is a confirmation process. This is either the LCR coordination process or a BFT confirmation process. Those processes have important different qualities that impact the potential benefits that might arise from a permissionless network. Nonetheless, in each case, nodes incur a cost in participating in the process of confirming a block. These costs mean that real resources might be expended in operating such networks securely or, alternatively, restrictions on the ability of nodes to freely trade within the network are required.

In economic terms, some trade-offs between permissioned and permissionless systems mimic the trade-offs between monopoly and competition. These involve broad costs. Competition tends to allow more options to ensure that the network protocol itself is as efficient as possible in serving users' needs. However, compared with a monopoly, it can also involve higher (production) costs. Similar considerations arise between permissioned and permissionless approaches.

4.1   BITCOIN PROOF OF WORK

As the permissionless innovation in blockchains came with Bitcoin, to consider the trade-off with permissioned environments, it is useful to focus first on the Proof of Work protocol that underlies Bitcoin. The notion of Proof of Work came from the research of Dwork and Naor (1992), who examined how adding costs to the sending of email may reduce spammers whose strategy was to flood networks with messages. For Nakamoto (2008), requiring nodes to incur costs to be considered to assemble, process and confirm blocks was a way of preventing malicious nodes from flooding a permissionless network.

In Bitcoin, the leader is chosen via a computational contest.[1] The seed for that contest is a hash of the previous block preceded by K zeros. Recall that a hash is an output of a function that has certain specific information as an input. Each participating node is given the challenge of racing to find that input for the hash $+K$. This is a solvable problem, and, indeed, K is set so that, on average, the pool of participating nodes will find a solution in 10 minutes. In practice, this means that K adjusts as more or less computational resources are brought by nodes to the contest.

The way to solve for the hash $+K$ is via brute force. Nodes invest in computing resources in order to conduct brute force trials more efficiently. The probability, p_i, that a node, i, with hash power of x_i, wins the contest for any given block is $\frac{x_i}{X}$ where $X = \sum_{j=1}^{n} x_j$. Thus, a node can increase its chances of winning a contest by investing in more computational capacity. The costs of this capacity are $c(x_i)$ (a non-decreasing and convex function) and include not only the amortised computer hardware costs but also the electricity costs in running computations (which can be considerable).

Interestingly, when the contest for a new block begins and before it is concluded, the arrival rate of a solution to the hash $+K$ problem follows a Poisson distribution similar to many models of R&D racing in economics (e.g., Reinganum, 1989). Here the prize, rather than a patent, is a block

---

[1] The origin for this idea was Back et al. (2002).

reward plus any transaction fees. The block reward, R, in Bitcoin involves receiving a specified number of bitcoins plus any transaction fees paid by agents writing transactions to that block.[2] A full exploration of this model and proof of its equilibrium existence (including for the endogenous adjustment of the difficulty of the problem) is contained in Ma et al. (2018). For our purposes here, it suffices to note that each node will solve the following maximisation problem each period:

$$\max_{x_i} p_i R - c(x_i)$$

They will do so simultaneously. This type of game is called a Tullock (2001) contest. For instance, if $c(x_i) = x_i$ for all i, then for a given number of nodes, n, the equilibrium hash power of each node will be $\hat{x}_i = \frac{n-1}{n^2} R$. Note that this is increasing in n and R. That is, an increase in the number of nodes competing for the reward increases the computational efforts of each node. Substituting this into the payoff function gives $\frac{1}{n^2} R$.

Suppose that there are sunk entry costs of C for each node. In the long-run, with symmetric nodes, $\hat{x}_i(n) = \hat{x}(n)$ and $p_i = \frac{1}{n}$ for all i and so expected profits given n, become $\frac{1}{n} R - \hat{x}(n) - C$. The number of nodes, n, will adjust to push expected profits to zero so that $\hat{n} = \sqrt{\frac{R}{C}}$. The equilibrium hash power becomes $n\hat{x}(n) = \frac{n-1}{n} R$. Thus, as shown more generally by Ma et al. (2018), the total cost in the blockchain is increasing in n, and if n could be restricted to one node, this would be trivially the lowest cost outcome. Of course, the benefits of reducing n would have to be compared to the costs in terms of blockchain consensus outcomes, something that will be discussed below.

4.2   PERMISSIONLESS LEADER SELECTION

As has been shown, the Bitcoin Proof of Work computational game is a Tullock contest where the amount of computational power applied by a node proportionately impacts its likelihood of winning the contest. At this stage, it is useful to note here that this rule for selecting a leader

---

[2] Huberman et al. (2021) provide an analysis of the economics of Bitcoin fee transactions and competition.

is actually the only structure of rule that would satisfy certain desirable properties of permissionless networks.

Specifically, Leshno and Strack (2020) demonstrate that for any given application of resources by nodes, that is, $\{x_1, ..., x_i, ..., x_n\}$, setting $p_i = \frac{x_i}{\sum_{j=1}^{n} x_j}$ for all i is the only structure for p_i that satisfies the following properties:

1. *Anonymity*: if any two nodes can change their identities, they inherit the selection probability of one another. Importantly, it does not allow the protocol to condition the history of the node's behaviour.
2. *Robustness to Sybil Attacks*: a node cannot split its performance into two or more entities and pose as a new entrant to increase its selection probability. Incidentally, this condition ensures that free entry is possible, and insiders cannot undertake certain actions that prevent others from entering.
3. *Robust to Merging*: nodes cannot increase their selection probability by merging. In other words, a permissionless network must forestall any incumbent advantages to effectively ensure that anyone can participate on equal terms.

Of these conditions, anonymity is most obviously desirable as the whole notion of a permissionless network is that anyone would be able to participate in the network and, therefore, no selection rule for the leader should be contingent on any particular agent.

The other two properties are desirable in allowing competition to persist on a level playing field amongst nodes. A Sybil attack is a situation where an adversarial agent creates a large number of pseudonymous identities to gain power or otherwise control a network.[3] The most familiar of these is the creation of new email accounts to send spam emails that may overcome spam filters. From the perspective of blockchain consensus, if a protocol is vulnerable to a Sybil attack, then there is no limit on the number of faulty nodes that might be created. In a permissionless environment, therefore, the BFT protocols, based on their being a known upper bound to the number of faulty nodes, will not be Sybil resistant. From an economic perspective, this property implies that there will be free

---

[3] The term 'Sybil' comes from a book of the same name about someone with dissociative identity disorder; a term that itself is no longer used in psychiatry.

entry by agents to establish nodes in the network; that is, if there are n nodes, a new entrant can come in and become node $n+1$. The final property ensures that there is no incentive for existing nodes to merge their computational resources to gain an advantage in the leadership contest. If this was not present, networks would not remain decentralised and would create pressures towards centralisation.

Leshno and Strack (2020) prove that the proportional selection rule for p_i as implied by the Bitcoin Proof of Work contest is the *only* rule that satisfies all of these properties.[4] Indeed, the properties do not imply one another, and so are independent conditions. Interestingly, achieving these properties is potentially quite difficult in practice. For instance, if node operators were risk averse, then they would have an incentive to merge and pool their resources. Such pools have become common in Bitcoin (see Cong et al., 2021). This suggests that when there is risk aversion, such pools cannot be avoided, and the blockchain network will not be as decentralised as might be considered desirable.

4.3   ATTACKS ON PERMISSIONLESS BLOCKCHAINS

If achieving consensus in a permissionless environment is necessarily costly, it is worth considering what the benefits might be. The usual benefit discussed is with regard to security. LCR protocols, as we have already seen, reach consensus quickly (with a simple leader message broadcast to the network) but leave open whether the leader might be malicious. The question, therefore, is whether having a permissionless network reduces the probability that a malicious leader might be chosen. Thus, the benefits of being permissionless hinge on the issue of transaction safety.

Bakos and Halaburda (2021) develop a simple model to examine these benefits. We have already noted that the cost of computing power used under a Proof of Work protocol is $nc(\hat{x}(n))$ (that we will write as nc for this section) that is deployed in a contest for a block reward of R. Free entry—a hallmark of permissionless networks—implies that $nc = R$. It has also been noted that with an LCR protocol, individual users will make a choice, k, of how many blocks they would want to be confirmed after the block with a relevant transaction before considering the payment final.

---

[4] Other selection rules may be based on nodes with $p_i = \frac{1}{n}$ or 'winner take all' with the node with the highest x_i always being selected as the leader.

What about an attacker? What are their incentives? Building on an analysis by Budish (2022); Bakos and Halaburda (2021) suppose that an attacker could earn a payoff of V_{attack} if their attack is successful. If the costs of an attack exceed V_{attack}, the attack will not take place. Otherwise, the network is vulnerable.

But what are the costs of an attack? Consider a blockchain that has been operating up to time t. At that time, a block, B_t, is proposed and confirmed to the blockchain. Absent an attack, at time $t + k$, an agent will consider the transaction at t to be final. In Bitcoin, that transaction is a payment of some quantity of bitcoins from one agent to another; say, Ariel to Bailey. As already discussed in Chapter 2, once a transaction is final (at $t + k$), the payee, Bailey, will perform their side of the contract by providing goods or services to Ariel. Suppose that payment has value, V_{attack}, to Ariel. Then if, somehow, the transaction in block B_t was removed and a new block B_t' confirmed, the payment value, V_{attack} of bitcoins would remain in Ariel's account. They could then spend those tokens again.

The Costs of a Double-Spend Attack

How could Ariel achieve this double-spend? Nakamoto (2008) realised that Ariel would have to be selected as the leader in an alternative but private (at least until $t + k$) blockchain. Ariel could achieve this by forking the blockchain at time t with blocks B_{t-1} and earlier being the same on both chains and blocks after $t+1$ being the same package of transactions as well. The only difference would be that B_t', with the relevant transaction removed, would be substituted for B_t. However, as blocks are hashed together, the hash for B_t' is different from B_t, and so the hashes of blocks up to and including B_{t+k} will also be different. Ariel will have to construct the whole chain.

That means putting in the work to solve the relevant computational puzzle and incurring computation costs in the process. For Ariel to have their fork adopted at $t+k$, their private chain would, by the LCR, have to be longer than the main blockchain at that time. While Ariel faces no in-chain competition, if Ariel does not match or exceed the computational resources employed by all other nodes working on the main chain, Ariel's fork will not be adopted, and the attack will fail. We already calculated that the computational resources being devoted to the main chain had a cost of nc per block, and so Ariel's probability of having the longest chain would be in relation to these costs.

It is commonly argued that a simple 51 per cent majority of nodes is all Ariel would need to mount a successful attack. If A was a parameter capturing Ariel's advantage in nodes—with $A = 1$ when Ariel has n nodes and $A > 1$ when Ariel has more than n nodes—a 51 per cent attack would have $A = \frac{51}{49} \approx 1.04$. However, this would only mean that Ariel would only be more likely to have the longest chain after k periods. Ariel may, therefore, choose to acquire more or less than 51 per cent of all nodes when mounting an attack.

Budish (2022) has calculated the expected duration, $t(A, k)$, of an attack that lasts at least $k + 1$ periods given a chosen advantage in terms of computational resources (A).[5] Suppose that $B_H(s)$ $(B_A(s))$ is the number of blocks confirmed on the honest (attacking) chain at time, s, then:

$$t(A, k) = E[\inf\{s : B_H(s) \geq 1 + k, B_A(s) > B_H(s)\}]$$

That is, $t(A, k)$ is the time at which an attack is complete creating a chain that has more blocks than the waiting period $1 + k$ and more blocks than the honest chain. This has the following closed-form solution:

$$t(A, k) = (1 + k) + \left[\sum_{i=0}^{1+k} \left(\frac{i+1}{A-1} \right) \cdot \frac{(1+2k-i)!}{(1+k-i)!k!} \left(\frac{A}{1+A} \right)^{1+k-i} \left(\frac{1}{1+A} \right)^{1+k} \right]$$

$t(A, k)$ is increasing in both arguments. Thus, the expected costs of an attack are $Anct(A, k)$.

Interestingly, for an attacker like Ariel, if the attack is successful, they will receive *all* of the block rewards created during $t(A, k)$. This is because this is an outside attack where the attacker has more than double the hash power in the system while computational difficulty has remained unchanged. The attacker receives all of the rewards because their forked chain is kept private until it has exceeded k periods or becomes the longest chain (whichever comes second). Thus, as Ariel is the sole agent building this fork, they would be selected as the leader for each block created and so be guaranteed all of the block rewards. This will offset the direct costs of the attack. $At(A, k)$ blocks will be expected on the attack chain compared with $t(A, k)$ on the main (or honest) chain.

---

[5] See also Chiu and Koeppl (2017).

However, this does not mean that Ariel's expected block reward is R for each block. Recall that the block reward is composed of newly minted tokens (which we can denote by θ) and transaction fees (which we can denote by η per period). Each block created by the attacker gives them new tokens, so the expected number of tokens would be $At(A, k)\theta$. Transaction fees are determined by the number of transactions created in $t(A, k)$ periods. This volume is unlikely to be influenced by the computational resources the attacker devotes to their alternative chain. Instead, it will mirror the number of transactions generated by users of the network during the attack period. Thus, for that part of the block reward, during the attack, the attacker receives $t(A, k)\eta$; the same amount as generated by the honest chain.[6] Thus, the total expected block reward to the attacker is $(A\theta + \eta)t(A, k)$.[7]

Given this, the net cost is:

$$Anct(A, k) - (A\theta + \eta)t(A, k)$$

Recalling that free entry implies $nc = \theta + \eta$, this net cost becomes:

$$(A - 1)\eta t(A, k)$$

If block rewards were predominantly newly minted tokens, then this net cost can become very close to zero.[8] This remarkable property arises because the permissionless nature of the network implies that computational costs are covered by expected rewards. This holds both for honest nodes working on an honest chain and an attacker working on a separate chain. It also suggests that such a permissionless network would

---

[6] There are some caveats here. First, the attacker is removing at least one transaction and maybe more from the blockchain, which also means the associated fees would not be paid. Second, it may be that the processing capacity of the network has reached capacity during the attack period, in which case it may be possible for the attacker to process more transactions.

[7] In Gans and Gandal (2021), and Halaburda et al. (2022) it was assumed that during an attack, the attacker confirmed only t blocks. This carried on an assumption made in an earlier version of Budish (2022). This might be an appropriate assumption if all or most of the block reward was comprised of transaction fees but not if a significant share is comprised of newly minted tokens.

[8] A related result arises in models that posit attacks taking the form of selfish mining— just to receive block rewards rather than double-spend. See Eyal and Sirer (2018) shows that an attacker with more than one-third of all hashing power could successfully mount such an attack.

be vulnerable to attacks even when the potential benefit to the attacker, V_{attack}, is low.

In practice, there are additional potential reasons why an attacker may face a significant positive net cost associated with an attack.

1. *Relative computational costs*: Budish (2022) makes the assumption that an attacker might have higher relative costs of computation than honest nodes; facing a cost inefficiency that adds to the attacker's costs of $\kappa \geq 1$. In this case, the net costs of an attack become $(1+\kappa)Anct(A, k) - (A\theta + \eta)t(A, k) = (\kappa A(\theta + \eta) + (A-1)\eta)t(A, k)$. Thus, for an attack not to take place requires that $V_{attack} < (\kappa A(\theta + \eta) + (A-1)\eta)t(A, k)$.

2. *Depreciation of token value*: Budish (2022) and Moroz et al. (2020) suppose that token value might depreciate if a double-spend attack were discovered even if the attack were successful because confidence in the network could be reduced. Thus, suppose that the value of tokens depreciates by Δ_{attack}. Block rewards (both tokens and fees) are denominated in token value as is, for the case of a double-spend attack, V_{attack}. Thus, if it is expected that the double-spend attack would be noticed and caused the token value to decline by Δ_{attack}, net costs become $Anct(A, k) - (1 - \Delta_{attack})(A\theta + \eta)t(A, k) = (\Delta_{attack}A\theta + (A - 1 + \Delta_{attack})\eta)t(A, k)$. Thus, for an attack not to take place requires that $V_{attack} < \frac{(\Delta_{attack}A\theta + (A-1+\Delta_{attack})\eta)t(A,k)}{1-\Delta_{attack}}$.

3. *Detection and roll-back*: Bakos and Halaburda (2021) explore the possibility that a double-spend attack could be detected and rolled back.[9] This means that the attacker would be left without V_{attack} and the expected block rewards $(A\theta + \eta)t(A, k)$. Suppose that detection and roll-back occurred with probability d. Then the attacker's expected return is $(1-d)V_{attack} - Anct(A, k) + (1-d)(A\theta + \eta)t(A, k)$ which sets the condition for there not to be an attack to: $V_{attack} < \frac{dA\theta + (A-1+d)\eta}{1-d}t(A, k)$.

4. *Detection and limited roll-back*: Suppose that the double-spend attack is detected and the blockchain restored to the original chain

---

[9] A more complex counter-attack by the agent who is harmed in the double-spend attack is explored by Moroz et al. (2020).

but that the attacker has been able to receive the value, V_{attack}, as this occurred off-chain and cannot be restored. In this case, the attacker loses the expected block rewards $(A\theta+\eta)t(A,k)$, only. Again using a detection probability of d, the attacker's expected return is $V_{attack}-Anct(A,k)+(1-d)(A\theta+\eta)t(A,k)$ which sets the condition for there not to be an attack to: $V_{attack} < (dA\theta+(A-1+d)\eta)t(A,k)$.

A final area where the costs of an attack might increase is where the capital required for computational resources is specific to mining. Thus far, it has been assumed that an attack requires the attacker to rent the relevant capital for the duration of the attack. However, for some blockchains, in particular, for Bitcoin, the computational best practice has become ASIC chips that are specialised in bitcoin mining. This complicates the analysis of an attack somewhat. For instance, if the attacker purchases equipment to mount an attack and build the longest chain after the attack, the free entry condition means that the equipment is unlikely to continue to earn sufficient revenues to cover costs. Alternatively, the attacker may take control of existing nodes and use them for the attack. In this case, assuming no further entry, the attacker also needs fewer computational resources for the attack as they would have to control over half the existing network. However, in this case, the future returns for that equipment rely on the continued high value placed on the network (allowing the tokens to maintain their value). Without this, a depreciation in the currency (Δ_{attack}) may increase costs further. The bottom line is that sunk capital costs associated with computation raise the costs of an attack.

4.4   Comparison with Permissioned Network

In Chapter 3, the BFT protocols examined were for permissioned networks where (a) the number of validating nodes (n) was known and (b) their identity was known. This is the hallmark of a permissioned network with the challenge of achieving consensus amongst known but distributed nodes. As was shown in that chapter, the functioning of these networks required the number of faulty nodes (or malicious ones) to be below some threshold. The threshold depended on the protocol being followed.

Transaction Safety in a Permissioned Network

While that chapter focussed on the ability to achieve consensus, there may still have been concerns that, if there was a significant and widespread attack, transactions might not be safe. The implicit assumption, however, was that the attack could be identified and mitigated after the fact to both counter incentives for the attack and to give users confidence in network security.

What is left to explore is what happens when attacks cannot be easily detected and, moreover, might be difficult to sanction. Bakos and Halaburda (2021) provide a simple model that captures these effects. At the heart of their model is an assumption that if an attack is detected (which happens with probability d) and if a node can be identified as being responsible (individually or as part of a coalition of other nodes), which happens with probability ρ, then nodes can be sanctioned with an imposed penalty of Φ. Φ is an exogenously given limit but may be regarded as the limit of penalties imposed by the legal system. In Chapter 5, it will also be shown how stakes might serve a non-judicially imposed penalty in Proof of Stake networks.

To facilitate comparison with permissionless blockchains, it is supposed that each block confirmed generates η in transaction fees. This accrues to the node confirming the block.

Let's suppose that a simple majority of nodes (i.e., $\underline{n} = \frac{n+1}{2}$) is sufficient to successfully engage in a double-spend attack by allowing what might be otherwise an invalid transaction to be confirmed to a block (say, by rewriting history to exclude a previously valid transaction). It is assumed that because every node knows the identity of other nodes, they can collude in this way. In addition, it is assumed that the attack requires re-confirming $k_p + 1$ blocks where k_p is the number of periods a user waits for block finality in this permissioned network.

If the joint value to the attacking coalition from this attack is V_{attack}, then they will not attack if:

$$V_{attack} + (1 - d)\eta(k_p + 1) < d\rho\underline{n}\Phi.$$

Note that this assumes that if an attack succeeds, the attacking coalition can share the block transaction fees between them but if the attack is detected, those fees are forfeit.

Comparison

This now allows a comparison of the resiliency of permissionless and permissioned blockchains with regard to transaction safety. In particular, the following can be defined where the 'detection and limited roll-back' case is applied for each:

$$V_p = d\rho\underline{n}\Phi - (1 - d)\eta(k_p + 1)$$

$$V_{pl} = (dA\theta + (A - 1 + d)\eta)t(A, k_{pl})$$

If V_{attack} is less than V_p, the permissioned blockchain is resilient and if it is less than V_{pl}, the permissionless blockchain is resilient. This means that a permissionless blockchain will be more resilient than a permissioned blockchain if $V_{pl} > V_p$ or:

$$(dA\theta + (A - 1 + d)\eta)t(A, k_{pl}) > d\rho\underline{n}\Phi - (1 - d)\eta(k_p + 1)$$

$$\implies ((A - 1 + d)t(A, k_{pl}) + (1 - d)(k_p + 1))\eta > d(\rho\underline{n}\Phi - A\theta t(A, k_{pl}))$$

Note that if transaction fees, η, are higher, the relative resiliency of the permissionless blockchain over the permissioned blockchain rises. In both environments, attacker costs are offset by the potential to earn transaction fees should the attack not be detected. However, in the permissionless network, higher transaction fees drive further entry by nodes which increases attacker costs; thereby, increasing resiliency.

This illustrates a general principle driving resilience. In both networks, greater numbers of nodes make attacks more costly. In a permissionless network, it is block rewards that create the incentives to enter. In a permissioned network, this is set by design. In a permissionless network, the per-node cost of those additional incentives is $\frac{\theta + \eta}{n}$, while in a permissioned network, it is $\rho\Phi$. Therefore, if it is difficult to raise $\rho\Phi$, a permissioned network may be less resilient than a permissioned network.

Cost Incidence

There is another key difference between permissioned and permissionless networks as identified by Bakos and Halaburda (2021). In permissioned networks, the costs ($\rho\Phi$) are not incurred in equilibrium. They are basically the threat that can be utilised by outside enforcement of blockchain behaviour by nodes. In permissionless networks, there may be no outside enforcement.[10] Thus, security is achieved by offering block rewards that incentivise good behaviour from nodes. Those block rewards are costly to the network (i.e., issuing tokens will depreciate the value of existing tokens) and so users are paying them upfront, as it were.

Both external enforcement triggered by bad behaviour and incentive alignment paid for by ongoing rewards can create safety. The threat of external enforcement is commonly examined in economics using the model of crime by Gary Becker (1968). While the use of ongoing rewards is a hallmark of models of efficiency wages that encourage effort by workers (Shapiro & Stiglitz, 1984). Thus, in trading off choices of permissioned versus permissionless blockchains, users will have to consider the benefits of resiliency that might be afforded by having a larger, permissionless network versus the benefits of lower operational costs that can be delivered in a permissioned environment.

4.5   Conclusion

The main economic difference between a permissionless and a permissioned network is that the former allows anybody to operate a node in the network while the latter restricts the set of nodes. As the challenge of securing a permissionless network is correspondingly higher than a network where all nodes are honest and trusted, the resource costs of permissionless security are correspondingly higher than those of permissioned networks. Of course, permissioned networks may be subject to attack in that malicious agents can take control of some nodes. To deter that, permissioned networks need to invest in ways of detecting problems and limiting the reward from attacks after the fact.

In reality, as Budish (2022) discusses, the costs of securing a permissionless network against hypothetical double-spend attacks are potentially quite costly because attackers can take advantage of participation constraints during attacks. Moreover, it is unclear whether double-spending, even though it is theoretically possible, will ever generate payoffs for attackers that make it worthwhile; especially for more liquid cryptocurrencies. Double-spending requires an attacker to be involved in a sizeable transaction with another party. Even if a double-spend attack

---

[10] There may be outside enforcement if, for instance, in a double-spend attack, the relevant contracts that are broken end up in a non-blockchain judicial process.

occurred, it would be known, at least to that other party, who is likely to have some recourse through traditional legal institutions.

All this favours permissioned networks as being more likely to be economically efficient. This is at least with respect to Proof of Work. As will be shown in the next chapter, Proof of Stake offers an opportunity to secure a network with fewer real resources than Proof of Work and hence, a greater claim to economic efficiency.[11]

Key insights from Chapter 4

- Under permissioned blockchains, a consensus is achieved using the fact that the number of nodes is known. For permissionless blockchains, the number of nodes is unknown. In Bitcoin that led Nakamoto (2008) to invent the longest chain rule for consensus.
- To limit Sybil attacks, permissionless blockchains require there to be costs to nodes from having validating and other roles on the network. To counter this, rewards must be given. These rewards are a mixture of block rewards (newly minted tokens) and transaction fees. They are set so that nodes facing costs will choose to participate in the network.
- Permissionless networks are subject to potential attacks by malicious nodes. Those agents, however, need to amass resources to create a significant (usually 51 per cent or greater) of total resources on the network. However, those costs are mitigated by the ability to earn rewards during an attack.
- To guard against such attacks while also encouraging participation, permissionless blockchains have to pay for security upfront (effectively with each transaction). By contrast, permissioned blockchains can protect networks by applying resources ex post to deter attackers.

---

[11] Brandenburger and Steverson (2021) note that under Proof of Work, when a computational puzzle is solved quickly, agents can get an estimate of how many other nodes are on the network—that is, a proof of presence. They show how this information can be used to amend Proof of Work protocols to achieve the same level of security at a lower resource cost.

References

Back, A., et al. (2002). Hashcash-a denial of service counter-measure.

Bakos, Y., & Halaburda, H. (2021). *Tradeoffs in permissioned vs permissionless blockchains: Trust and performance.* NYU Stern School of Business working paper.

Becker, G. S. (1968). Crime and punishment: An economic approach. In *The economic dimensions of crime* (pp. 13–68). Springer.

Brandenburger, A., & Steverson, K. (2021). *Using 'proof-of-presence' to coordinate.* Technical report.

Budish, E. B. (2022). *The economic limits of bitcoin and anonymous, decentralized trust on the blockchain.* University of Chicago, Becker Friedman Institute for Economics Working Paper No. 83.

Chiu, J., & Koeppl, T. V. (2017). *The economics of cryptocurrencies—Bitcoin and beyond.* SSRN 3048124

Cong, L. W., He, Z., & Li, J. (2021). Decentralized mining in centralized pools. *The Review of Financial Studies, 34*(3), 1191–1235.

Dwork, C., & Naor, M. (1992). Pricing via processing or combatting junk mail. In *Annual international cryptology conference* (pp. 139–147). Springer.

Eyal, I., & Sirer, E. G. (2018). Majority is not enough: Bitcoin mining is vulnerable. *Communications of the ACM, 61*(7), 95–102.

Gans, J. S., & Gandal, N. (2021). Consensus mechanisms for the blockchain. In *The Palgrave handbook of technological finance* (pp. 269–286). Springer.

Halaburda, H., Haeringer, G., Gans, J., & Gandal, N. (2022). The microeconomics of cryptocurrencies. *Journal of Economic Literature, 60*(3), 971–1013.

Huberman, G., Leshno, J. D., & Moallemi, C. (2021). Monopoly without a monopolist: An economic analysis of the bitcoin payment system. *The Review of Economic Studies, 88*(6), 3011–3040.

Leshno, J. D., & Strack, P. (2020). Bitcoin: An axiomatic approach and an impossibility theorem. *American Economic Review: Insights, 2*(3), 269–86.

Ma, J., Gans, J. S., & Tourky, R. (2018). *Market structure in bitcoin mining.* Technical report, National Bureau of Economic Research.

Moroz, D. J., Aronoff, D. J., Narula, N., & Parkes, D. C. (2020). *Double-spend counterattacks: Threat of retaliation in proof-of-work systems.* arXiv preprint arXiv:2002.10736

Nakamoto, S. (2008). Bitcoin: A peer-to-peer electronic cash system. *Decentralized Business Review,* 21260.

Reinganum, J. F. (1989). The timing of innovation: Research, development, and diffusion. *Handbook of Industrial Organization, 1,* 849–908.

Shapiro, C., & Stiglitz, J. E. (1984). Equilibrium unemployment as a worker discipline device. *The American Economic Review, 74*(3), 433–444.

Tullock, G. (2001). Efficient rent seeking. In *Efficient rent-seeking* (pp. 3–16). Springer.

CHAPTER 5

Proof of Work Versus Proof of Stake

Abstract This chapter discusses the differences between Proof of Work (PoW) and Proof of Stake (PoS) consensus mechanisms in blockchain networks. PoW, as in Bitcoin, requires significant resources, resulting in high energy consumption. PoS, a more energy-efficient alternative, requires nodes to hold and commit tokens as stakes. This chapter compares the security of both PoS and PoW, highlighting key trade-offs. While PoS can potentially have greater attack detection and response capabilities, it generally requires more off-chain coordination. Understanding these trade-offs is vital for designing efficient and secure blockchain networks.

Keywords Proof of Work · Proof of Stake · blockchain · Security · Consensus mechanisms

Work is costly as it requires the employment and use of real resources. In the case of Bitcoin, these resources were computational hardware and energy to power that hardware. The energy use alone for the entire network was estimated to be equivalent to that of a small country like Sweden (De Vries 2018). Moreover, as the value of bitcoins rose, it would attract and utilise even more resources. That, of course, was by design. As

---

J. Gans, *The Economics of Blockchain Consensus*,
https://doi.org/10.1007/978-3-031-33083-4_5

was shown in the previous chapter, it is precisely because resource use is costly that it makes attacks on the network costly.

Given this resource utilisation arising from Proof of Work protocols, there has been considerable research undertaken in computer science to find alternatives that can achieve Sybil resistance in a permissionless blockchain without the use of real resources. The most prominent alternative has been called Proof of Stake.

Under Proof of Stake, nodes that wish to participate in the network, either by proposing blocks or by validating them, are required to hold a certain amount of tokens and commit not to withdraw them while consensus is achieved. As will be shown, this relies on token stakes being provided and broadcast at the start of any round to achieve consensus. This is in contrast to Proof of Work, where no such commitment is required of nodes. This raises the issue of whether Proof of Stake protocols, which require very little energy use, can be as secure as Proof of Work.

As Proof of Work has already been outlined in the previous chapter, the first task here is to describe how Proof of Stake operates in a permissionless setting. Then, using approaches similar to the previous chapter, the security of such networks will be evaluated and compared to the security of Proof of Work. Once again, the goal is to identify key trade-offs in blockchain design.

5.1   Proof of Stake
in a Permissionless Environment

One of the characteristics of a permissionless Proof of Work protocol is that no one needs to know who or how many are participating in the contest to become the leader in any given time period. That node just emerges, and because of the nature of the computational contest, it is possible to presume that each node is likely to be selected in proportion to their overall share of resources deployed.

For Proof of Stake, the equivalent of resources is the share of staked tokens nodes have that could determine their selection as the leader to propose a block. The problem is that the protocol has to know what the pool of staked tokens is in order to select one. That means that staking must take place *prior* to selection. And moreover, if the stake is going to matter—that is, to give nodes an incentive to perform well and propose blocks honestly—that means the stake must be maintained throughout the consensus process. For example, if a node's token was selected, but then the node went silent, no block would be proposed, and consensus could not be achieved. This means that the Proof of Stake environment has to provide an incentive for validating nodes to perform their functions. As will be shown, some of this incentive comes from the block rewards—notably transaction fees—that staking nodes can earn but also, given the possibility of attacks, some off-chain punishments.

In what follows, the protocols for Proof of Stake blockchains are outlined and analysed. While the Longest Chain Rule (LCR) approach, which was developed to handle a permissionless setting, can be modified for Proof of Stake, after considering that approach, we turn to look at how the BFT approach has been modified for a permissionless environment and is now the predominant way Proof of Stake protocols are implemented for blockchains.[1]

A Longest Chain Rule Approach

The most natural starting point for considering a Proof of Stake mechanism in a permissionless environment was to draw inspiration from Nakamoto's coordination mechanism: a leader is selected randomly to propose a block, and forks are resolved by nodes being programmed to extend the longest chain. The first Proof of Stake network, Peercoin (King & Nadal, 2012), took precisely this approach.

While in Proof of Work, the length of a 'round' is determined by how long it takes for someone to win the computational contest, under Proof of Stake, as there is no contest. Time is simply specified as discrete time slots. Either a block is proposed for that time slot or it is not. If it is proposed, then the chain would be extended. If not, the chain length would be unchanged.

In practice, there is a chance no block would be proposed. Recall that a leader is selected by choosing a particular token at random and offering its owner the opportunity to propose a block. If the owner is not prepared to perform the function of assembling messages into a block, then time passes and a new round would begin. There is, however, an incentive to be available to be selected as a leader as they would receive a block reward and transaction fees. Note that this selection method for the leader

---

[1] The possibility of 'zero cost' attacks is explored further in Gans and Halaburda (2023).

satisfies the Leshno and Strack (2020) conditions as an agent's likelihood of being selected was in proportion to their token holdings.

To see this, suppose there were ten nodes that each held one of the ten already issued tokens in a blockchain. The probability that anyone is selected to propose the next block is $\frac{1}{10}$. Suppose that if they do so, another token is issued and allocated to them. This means that in the next round, the previous leader now has a $\frac{2}{11}$ chance of being selected again, which is greater than the likelihood of $\frac{1}{11}$ for others. Precisely because a Proof of Stake network is bootstrapped on its own tokens, there is a tendency towards concentration of power.[2]

A potentially more significant issue arises when there is a fork in the blockchain. Recall that under Proof of Work, each forked branch of the blockchain involves a different computational contest and thus, any node with computational resources must choose which branch to work on. A node could work on both, but this would reduce their chances of winning the contest on either. Thus, nodes have an incentive to choose a branch, and the longest chain rule guides what honest nodes would do.

When a fork appears in a Proof of Stake protocol, it can be recommended that honest nodes work on one branch and the longest branch, if any. However, unlike Proof of Work, they do not have an incentive to pick and choose a single branch following a fork. Instead, they maximise their chances at a block reward by choosing to confirm a block should any opportunity—regardless of branch—arise. This has been called the *nothing at stake* problem. Saleh (2021) shows, however, that if having persistent forks reduces the demand by users for a blockchain's tokens, then honest nodes have an incentive to coordinate on a single branch—with the longest one providing a focal point if branches are of different lengths. Intuitively, the largest stakeholders have an incentive to coordinate.

This means that the mechanics of a double-spend attack is different than under Proof of Work. Consider an attacker who wants to remove a transaction at B_t. Under Proof of Stake, the attacker only has an opportunity to create a fork in the blockchain at a point where they, themselves, have been the leader, as nodes can only propose blocks if they hold the private key to the relevant public key of the selected leader. Suppose that

---

[2] Roşu and Saleh (2021) challenge this baseline prediction. They show that any tendency towards concentration peters out before there is, say, a single dominant token holder.

their most recent block before B_t where the attacker has been the leader is $B_{t-\alpha}$. (If the attacker, as we will assume, has a majority of the staked tokens, then α may be quite small.) The attacker will begin a new private chain at $B_{t-\alpha}$ and will be the only agent confirming blocks to that chain. Hence, they can substitute in their own censored B_t, as the opportunity arises.

To complete the attack, the attacker will have to devote all of their tokens to only appending blocks to the attack chain and not the main chain. Suppose there are n nodes each with a stake of s. Let S be the total number of staked tokens. The cost of holding a stake of tokens for a period is res where e is the exchange rate between tokens and fiat money and r is the prevailing rate of interest on fiat money. The attacker acquires AS (where $A(> 1)$ is as before) in token holdings to mount the attack; e.g., if $A = 1$, then the attacker matches the previous total of staked tokens.

It is important to note here that, unlike the case for an attack under Proof of Work, the attacker must control stakes that were already committed to the network. There are two reasons for this. First, there may be stakeholding periods that matter for whether you can be selected as a leader, and that selection must take place before any transactions associated with the attack. Second, if the attacker created new stakes in order to facilitate the attack, this has the effect of pushing the returns of honest nodes negative and causing exit. This would not in itself thwart an attack, but the sudden influx, being observed, may alert others that an attack is in progress.[3] The dynamics of that are potentially complex, so here, it is assumed that the attacker controls or is able to acquire control of existing staking nodes.

How long will the attack last? For the attack to be successful, it has to become the longest chain, in terms of blocks, and also to exceed the waiting period for a transaction to be considered final by the counterparty to the attacker in period t. For Proof of Work, mining the attacker's private chain could begin at time t, with the substitute block as the block following the common parent block of both the attacker's chain and the main chain. It would succeed if, at some time, τ, $B_H(\tau) \geq 1 + k$ and $B_A(\tau) > B_H(\tau)$. Note that because of the way the computational context

---

[3] Just as the move of computational resources to a private chain may alert others to an attack under Proof of Work.

works, k can represent both the number of time periods and the number of blocks before a transaction is considered final.

For Proof of Stake, success requires $B_A(\tau) > B_H(\tau)$ but is a transaction final k confirmed blocks on the honest chain or after k time periods. These two concepts are the same under Proof of Work but differ under Proof of Stake. During the attack, the rate of progress on both the attacker's and the honest chain are reduced, so it takes longer to confirm k blocks than it would wait k periods after t, the time and block where the transaction under attack took place.

Suppose that users use time rather than counting blocks to determine economic finality. In this case, attack success would involve the attacker making their chain public at the earliest point after k *periods* have passed, i.e., at $t + k$. Notice that this does not mean either the attacker's private chain or the main chain has to confirm k additional blocks. Just that the time is now beyond $t + k$.

Given this, the expected duration for an attack is given by:

$$t_{PoS}(A, k) = E[\inf\{\tau : \tau \geq 1 + k, B_A(\tau) > B_H(\tau)\}]$$

Note that in this definition, $1 + k$ periods must pass, and then the attack chain must become longer than the main chain.[4] This latter component would include the time taken to create the α block deficit unless the attacker plans the attack at $t - \alpha$. Thus, the cost of an attack is $AreSt_{PoS}(A, k)$. Once again, the attacker chooses A to minimise $AreSt_{PoS}(A, k)$.

As was the case with a double-spend attack with Proof of Work, the attacker also earns block rewards under Proof of Stake with Nakamoto consensus. The attacker expects to receive $\frac{A}{A+1}t_{PoS}(A, k)eR$ where R is the number of tokens from new token rewards and also transaction fees. Note that, unlike the Proof of Work case, because time periods are fixed, there is no difference between token rewards and transaction fees during an attack. Thus, the net cost of an attack is:

$$t_{PoS}(A, k)res_{attack} - \frac{A}{A + 1}t_{PoS}(A, k)eR$$

---

[4] If finality were measured in blocks rather than time, this becomes: $t_{PoS}(A, k) = E[\inf\{\tau : B_H(\tau) \geq 1 + k, B_A(\tau) > B_H(\tau)\}]$. Thus, a different t_{PoS} would need to be used, but otherwise, the analysis is unchanged.

The fact that an attack under Proof of Stake comes from internal stakes, has implications for the examination of the free entry condition. As the attacker's stake, $s_{attack} = \frac{A}{A+1}S$, were part of the total staked tokens, S, prior to the attack, the free entry condition remains $R = rS$. Note that this condition would continue to hold during the attack, so there would be no entry or exit from honest nodes.

Given this, we can demonstrate the following:

Proposition 4 In a Proof of Stake blockchain with Nakamoto consensus, the net cost of an attack is always 0.

This can be seen by substituting the free entry condition into the attacker's cost, which gives a net attack cost of:

$$\frac{A}{A+1}t_{PoS}(A,k)e(rS - R) = 0$$

There may be other factors that limit the attacker.[5] This calculation assumes, of course, that the attacker does not face any borrowing or liquidity constraints that limit the number of tokens they can acquire (Saleh, 2021) or increase the rate of interest, r, the attacker must pay.

Finally, for completeness, suppose that with probability d, this attack is detected, and the tokens earned as a block reward as well as its stake, are taken from the attacker. Then, the net costs of an attack become:

$$\frac{A}{A+1}deRt_{PoS}(A,k) + deAS = d\frac{A}{A+1}\left(t_{PoS}(A,k) + \frac{1}{r}\right)eR$$

A BFT Approach

As Proof of Stake evolved, computer science researchers began to examine whether a BFT approach could be used as the basis for blockchain consensus in a permissionless environment. A significant breakthrough on

---

[5] In Gans and Gandal (2021), it was assumed that the form of a double-spend attack was similar to Proof of Work in that a secret chain was created by the attacker and then revealed when it was longest. However, even in this case, the share of block rewards would not be 100 per cent of those during the attack period but instead proportional to A of the overall potential rewards. Thus, the net cost of an attack would still be positive. Saleh (2021) finds a positive cost of a double-spend attack because he assumes there are no block rewards.

this path was Tendermint Buchman (2016), Kwon (2014). The Tendermint protocol has many variations, but there are three key elements:

1. Validating nodes would commit a stake of tokens with there being costs and penalties associated with a failure to perform validating functions. This would determine the number of nodes, n.
2. Consensus would be achieved in a single shot of BFT agreement consisting of two rounds of voting with a block being confirmed if $\frac{2}{3}$ of committed nodes vote to confirm the block.
3. A leader to propose a block is selected from the committed validating nodes in a round-robin. This means that everyone will know who the leader is in advance. However, if a leader is not performing (e.g., they are faulty), then a round will be restarted with a new leader quickly.

The details of how this protocol worked in practice involved various solutions that might arise when there were asynchronous time issues and need not concern us here. Importantly, validating nodes were identified by their digital signatures. That, combined with time periods being of a specified length, meant that this BFT protocol relied upon features to improve security that we previously discussed in Chapter 3.

An important feature of BFT protocols used to confirm blocks on the blockchain is that the stakes put up by validating nodes represent tokens that can be taken from their owners. Recall that this was not possible without their permission for normal tokens in a blockchain, as it would require their private key to facilitate the transfer. Validating nodes give up that security for their stakes. This, combined with the fact that nodes can be identified in terms of their behaviour, allows the BFT protocol to create the ability to punish nodes for bad behaviour by *slashing*.

Slashing is the removal of all or part of a validating node's tokens if, for instance, they fail to perform a designated role—e.g., leadership—because they are having connection difficulties or they are involved in voting to commit two different blocks committing a block that does not have $\frac{2}{3}$ of nodes agreeing to the block or pointing to the same past block, or voting on a block that points to different past blocks (Buterin, 2017). Importantly, for many of the attacks that may be attempted on a BFT blockchain, because actions are signed by a node with a slashable stake, then if that attack is detected, the nodes involved can have their entire stake removed. These strong slashing conditions mean that blocks can be considered economically final very quickly following their addition to a blockchain.

Specifically, suppose that validating nodes must wait a certain number of periods, comprising the number of blocks after they stake until they are selected, t_s, plus a number of blocks after being selected and proposing a block, t_p, before releasing their stake. The cost of this will be $A\frac{n}{2}(t_s + t_p)reS$ although they will also earn $A\frac{n}{2}t_peR$. With probability d, their attack is detected, in which case they lose $A\frac{n}{2}eS$ and any block rewards during the attack. As the length of time for the attack is fixed at $t_s + t_p$, the attacker need not commit more than 51 per cent of all nodes to attack the network, so $A \approx 1$. Thus, the overall expected return from an attack is:

$$V_{attack} - \frac{n}{2}(t_s + t_p)reS + (1-d)\frac{1}{2}t_peR - d\frac{n}{2}eS$$

To pin down the threshold on the attacker's payoff, V_{attack}, that would determine the resiliency of the network, we need to pair this expected return with the free entry condition for validating nodes. As it turns out, different BFT protocols give rise to different conditions.

For Tendermint, there is, at any given time, a fixed number, n, of validating nodes. An agent can become a validating node by proposing a stake that is one of the nth largest stakes. Thus, the free entry condition is:

$$\frac{1}{n}eR = re\hat{S}$$

For, say, Ethereum (Buterin & Griffith, 2017), the stakes are fixed at S, but the number of nodes is endogenous. Thus, the free entry condition is:

$$\frac{1}{\hat{n}}eR = reS$$

In each case, however, $R = nrS$ so it is the size of R that determines the total amount, nS staked. Thus, the resiliency threshold is any V_{attack} such that:

$$V_{attack} < \left(t_s + d(t_p + \frac{1}{r})\right)\frac{1}{2}eR$$

Notice that, unlike the case, for an LCR protocol, the net cost of an attack is expected to be positive. This would even occur if the probability the attack is detected, d 0. This perhaps explains why a BFT approach has become more popular in Proof of Stake blockchains.

5.2   COMPARISON WITH PROOF OF WORK

Proof of Work and Proof of Stakes are two methods whose intended goal is to make it more expensive to destroy a blockchain network than to operate it. At the same time, as these networks bring value to a wide distribution of agents in the economy, they are vulnerable to public goods problems where the incentives to maintain the network may be less than the private incentives for some agents to take actions that may compromise the long-term viability of the network.

In our evaluation of the security properties of Proof of Work and Proof of Stake, we have excluded the role that existing institutions may play in that security. There have been many instances where blockchains have engaged in hard forks to change the fundamental properties of a blockchain. This might be to change the size of blocks (Bitcoin versus Bitcoin Cash) all the way to change from Proof of Work to Proof of Stake (Ethereum 1.0 to Ethereum 2.0). It is safe to say that the consensus to achieve such change was not as a result of unilateral attacks by a single agent or coordinated attacks by a group of agents. Instead, it was a transparent community discussion and resolution that led the majority of participants to favour a new chain. Similarly, when there are attacks, it is possible that the existing institutions off-chain may resolve or roll-back the consequences of such attacks. Certainly, if someone had entered into a contract with someone who then engages in a double-spend attack, the aggrieved party will have recourse to the legal system to resolve that contractual dispute. As the value of the transaction involved grows, surely mitigation by existing social and legal institutions becomes more likely. This suggests that the concerns raised by Budish (2022) that blockchains become more vulnerable to catastrophic attack as they scale must be tempered by the role existing institutions can play for precisely that set of circumstances.

Thus, when evaluating Proof of Work versus Proof of Stake, our focus should be on what these security protocols do concerning the day-to-day operations of blockchains. For Proof of Work, the main costs to an attacker are not necessarily from the sunk costs expended to begin an attack nor the costs of redeploying computational resources elsewhere but are, instead, the costs associated with using those computation resources during the attack. In Chapter 4, we showed that the direct costs of an attack were $Anct_{PoW}(A, k)$, which implied that the scale of an attack would require computational resources being deployed to match those deployed by honest nodes. However, these are mitigated by the rewards an attacker may be able to earn during the attack $(A\theta + \eta)t_{PoW}(A, k)$ which are, by design, proportionate to the scale of computation resources devoted to the attack itself. Indeed, the block reward, $\theta + \eta$ tells us something about what the relationship between the cost and mitigating rewards from an attack might be. Specifically, free entry means that, for honest nodes, $nc = \theta + \eta$; i.e., they balance one another precisely because the block reward is the funding instrument for compensating those providing security for the network. Thus, on net, an attacker faces costs of $(A - 1)\eta t_{PoW}(A, k)$. Thus, rather than the costs of an attack being high relative to the costs of maintaining a Proof of Work network, they are virtually symmetric.

As was shown above, for Proof of Stake, the same symmetric outcome can arise. Indeed, as Proposition 4 shows, the net cost of an attack on an LCR protocol can plausibly be 0. This might be relaxed by imposing stakeholding periods—that is, a requirement to stake for a certain period of time before being able to validate a block and a requirement to leave the stake in place for another period of time. This makes it difficult to mount an attack based on new staking resources to the network but still leaves the network vulnerable if an attacker can acquire nodes with existing stakes (see also the Box: An economic framing of the trade-off).

An economic framing of the trade-off Abadi and Brunnermeier (2022) provide an economics approach to evaluating the trade-off between Proof of Work and Proof of Stake. They define a communication game whereby agents send messages from a permissible set (i.e., a vocabulary and also restrictions based on the time and the agent sending the message) at a cost of κ for each message sent to maximise a payoff that is the sum of ongoing utility from allocations, the utility of a terminal state and the cost of sending messages. If some agents are faulty, there is no ability for others to wait for agreement from every agent for a given transaction. Moreover, akin to the asynchronicity assumption, non-faulty agents cannot identify faulty ones and agents cannot tell the time at which messages are sent or received. Mirroring the computer science literature, Abadi and Brunnermeier (2022) look for equilibria that are Strong Nash Equilibria (that is, are robust to coordinated deviations by coalitions of agents).

They start from the assumption that the protocol is fault tolerant in that if a majority of agents are non-faulty, then those agents will have no incentive to deviate from the prescribed consensus protocol. However, as we saw in Chapter 3 when time is asynchronous, then a coalition of agents with greater than a third of all nodes can send different messages to distinct subgroups of other agents and effectively engage in double-spending—or more generally, persuading one group of agents that a set of transactions have been confirmed and the other group of agents that those same transactions have not been confirmed. The way to deter the use of such attacks is to ensure that attackers have ex ante costs of deviating from the protocol, e.g., by making messages costly to send or face ex post punishment once deviations are discovered, e.g., prevent attackers from engaging in future transactions with their own staked tokens.

Abadi and Brunnermeier (2022) show that in this case, either Proof of Work is required (involving real resources in sending messages, $\kappa > 0$) or Proof of Stake is required where the attacker receives a different terminal payoff depending upon whether they have been honest or not. If they are dishonest, that terminal payoff is reduced. Thus, if the utility from an attack V_{attack} is less than κ multiplied by the number of messages in an attack plus the loss in terminal utility from being dishonest, the network can be secure. What this shows is that if you want pure resource efficiency (no waste in the system, $\kappa = 0$ and pure allocative efficiency (there are no restrictions on otherwise feasible terminal states), you are out of luck. That said, as those costs are additive in providing a disincentive to attack, it is possible to use a combination of them to satisfy the attacker's incentive constraint not to attack. As message costs and restrictions on terminal utility can impact all agents, it is likely

> that some intermediate combination of Proof of Work and Proof of Stake would be optimal for any given blockchain.

What this all means is that there is relatively little difference between Proof of Work and Proof of Stake when it comes to security provisions in an LCR consensus protocol. Both, as Gans and Gandal (2021) show, create incentives to devote resources to security—one computational and the other financial—that are driven by block rewards. And both have rewards in tokens which create incentives for nodes not to allow network disruptions or attacks that may cause those tokens to depreciate in value. Finally, both rely on the off-chain mitigations that might arise to roll back the consequences of an attack to itself deter such attacks.

Instead, the focus turns to the type of mitigations that may be possible. It is here that Proof of Stake, by requiring tokens to be staked and then put at risk if bad behaviour is detected, may offer an advantage. By contrast, as noted by Buterin (2016), Proof of Stake breaks the symmetry between attack costs and operating costs

> ... by relying not on rewards for security, but rather penalties. Validators put money ("deposits") at stake, are rewarded slightly to compensate them for locking up their capital and maintaining nodes and taking extra precaution to ensure their private key safety, but the bulk of the cost of reverting transactions comes from penalties that are hundreds or thousands of times larger than the rewards that they got in the meantime. The 'one-sentence philosophy' of Proof of Stake is thus not 'security comes from burning energy,' but rather 'security comes from putting up economic value-at-loss.' A given block or state has \$X security if you can prove that achieving an equal level of finalisation for any conflicting block or state cannot be accomplished unless malicious nodes complicit in an attempt to make the switch pay \$X worth of in-protocol penalties.

In other words, the difference between Proof of Stake and Proof of Work comes from believes that d, the probability an attack is detected and punishments enacted, is potentially much higher for Proof of Stake with BFT consensus compared with Proof of Work LCR consensus. Again from Buterin (2016):

Theoretically, a majority collusion of validators may take over a Proof of Stake chain, and start acting maliciously. However, (i) through clever protocol design, their ability to earn extra profits through such manipulation can be limited as much as possible, and more importantly (ii) if they try to prevent new validators from joining, or execute 51% attacks, then the community can simply coordinate a hard fork and delete the offending validators' deposits. A successful attack may cost $50 million, but the process of cleaning up the consequences will not be that much more onerous than the geth/parity consensus failure of 2016.11.25. Two days later, the blockchain and community are back on track, attackers are $50 million poorer, and the rest of the community is likely richer since the attack will have caused the value of the token to go up due to the ensuing supply crunch. That's attack/defense asymmetry for you.

Interestingly, the best way to raise d turns out to have good off-chain coordination. Proof of Stake protocols are built around that idea, but, interestingly, as a result, the ability to fully automate blockchains is subverted. This illustrates why some of the most devoted blockchain adherents are sceptical of Proof of Stake approaches. As we will see, however, in Chapter 7, there is scope to use mechanism designs from economics to take advantage of the staking properties of Proof of Stake protocols to automate more of the mitigations that currently require off-chain efforts.

Key insights from Chapter 5
- In permissionless Proof of Stake, block leaders are chosen at random each round from the set of nodes that have committed stakes to the network. Under LCR coordination, a double-spend attack is possible by creating an alternative chain with a greater amount of staked tokens than the main chain.
- The net cost of an attack in permissionless Proof of Stake with LCR coordination is potential 0 because the direct costs of the attack (interest on staked tokens) are completely offset by expected block rewards during the attack. If an attack can be detected and punishments enacted, this can reduce the return to double-spend attacks as can concerns about token depreciation if an attack is detected.
- A BFT approach to permissionless Proof of Stake involves validating nodes committing to be available in each round and then blocks being confirmed if $\frac{2}{3}$ of nodes confirm a proposed block. If there is bad behaviour detected, the nodes in question can face slashing,

which burns some or all of their staked tokens. The security of the network critically depends on the ability to detect and punish attackers.

- The trade-off between Proof of Work and Proof of Stake depends on the relative detection probabilities and reaction to attacks which is arguably greater for Proof of Stake. However, Proof of Stake tends to require more off-chain coordination than Proof of Work.

REFERENCES

Abadi, J., & Brunnermeier, M. K. (2022). *Blockchain economics*. Technical report, Federal Reserve Bank of Philadelphia.

Buchman, E. (2016). *Tendermint: Byzantine fault tolerance in the age of blockchains* [Ph.D. thesis].

Budish, E. B. (2022). *The economic limits of bitcoin and anonymous, decentralized trust on the blockchain*. University of Chicago, Becker Friedman Institute for Economics Working Paper No. 83.

Buterin, V. (2016, December 30). A proof of stake design philosophy. *Medium*. https://medium.com/@VitalikButerin/a-proof-of-stake-design-philosophy-506585978d51

Buterin, V. (2017). Minimal slashing conditions. *Medium*.

Buterin, V., & Griffith, V. (2017). *Casper the friendly finality gadget*. arXiv preprint arXiv:1710.09437

De Vries, A. (2018). Bitcoin's growing energy problem. *Joule, 2*(5), 801–805.

Gans, J. S., & Gandal, N. (2021). Consensus mechanisms for the blockchain. In *The Palgrave handbook of technological finance* (pp. 269–286). Springer.

Gans, J. S., & Halaburda, H. (2023). *"Zero cost" double-spend attacks on permissionless blockchains*. Technical report.

King, S., & Nadal, S. (2012, August 19). *Ppcoin: Peer-to-peer crypto-currency with proof-of-stake*. Self-published paper (1).

Kwon, J. (2014). Tendermint: Consensus without mining. *Draft v. 0.6, fall 1*(11).

Leshno, J. D., & Strack, P. (2020). Bitcoin: An axiomatic approach and an impossibility theorem. *American Economic Review: Insights, 2*(3), 269–286.

Roşu, I., & Saleh, F. (2021). Evolution of shares in a proof-of-stake cryptocurrency. *Management Science, 67*(2), 661–672.

Saleh, F. (2021). Blockchain without waste: Proof-of-stake. *The Review of financial studies, 34*(3), 1156–1190.

CHAPTER 6

Cryptography Versus Incentives

Abstract This chapter explores the trade-off between using cryptography and incentives in blockchain systems, particularly in the context of smart contracts. Cryptography provides security and permissions, but can limit contract offerings and increase computational costs. Incentives encourage positive behavior, but require access to hidden information. By examining the example of front-running in blockchains with smart contracting layers, the chapter highlights the distinct solutions provided by cryptography and incentives. Ultimately, striking a balance between these two approaches is crucial for ensuring efficient and secure blockchain operations.

Keywords Cryptography · Incentives · Smart contracts · Front-running · Blockchain security

Blockchains are a consensus record of messages that have been confirmed by a distributed network. The messages themselves have to be valid. In particular, they often require a digital signature that is checked to see if the sender has permission to send the message and cause any consequent updating of ledger entries. It is cryptography that allows a large set of agents the ability to update records. Public key cryptography allows only those with permission to append a record (that is, those who hold the associated private key) to do so. That permission can be transferred. But

© The Author(s), under exclusive license to Springer Nature   85
Switzerland AG 2023
J. Gans, *The Economics of Blockchain Consensus*,
https://doi.org/10.1007/978-3-031-33083-4_6

importantly, cryptography also creates a layer of non-public information which means that the blockchain rules cannot depend on or change with that information. That can be beneficial in terms of security but, at the same time may limit the types of contracts that can be offered.

An alternative to a complete reliance on cryptography to assign permission and rights is using incentives. Incentives can drive good behaviour and punish bad behaviour. But to do so, they require access to information. If that information is hidden by cryptography, some incentives cannot be provided. Thus, there is a trade-off between the use of cryptography and the use of incentives to encourage certain behaviours.

This chapter will examine this trade-off using specific examples of smart contracting opportunities that have been proposed. It will show that contracting hazards appear in public blockchains and that both cryptography and incentives are potential—albeit distinct—solutions. We have seen variants of this trade-off in Chapter 3, where it was shown that digital signatures could aid in achieving blockchain consensus and, in their absence, ways of punishing bad behaviours must be created. Here the focus shifts to the trade-offs in message sending rather than confirmation per se. Specifically, the focus here will be on the example of front-running that has been pervasive in blockchains with smart contracting application layers, as this is an area that demonstrates the trade-off starkly.[1]

To forecast, the main trade-off between incentives and cryptography is that the latter involves additional computational costs that can slow down blockchain operations, while the former may take time to yield the desired results as an equilibrium outcome.

6.1   BLOCKCHAIN FRONT-RUNNING

Front-running has become a serious issue for smart contracts in blockchain ecosystems; threatening to completely undermine its potential.[2] The problem is straightforward. When a contract is placed on a blockchain such as Ethereum, there is a performance obligation on one party that, when achieved, triggers a payment in tokens from another party. Sometimes these contracts are open offers—such as a bounty or

---

[1] This chapter is based on and draws liberally from Gans and Holden (2023).

[2] See Catalini and Gans (2020), Gans (2022) and Holden and Malani (2021) for overviews of the economic potential of the blockchain to solve contracting issues.

reward. When performance occurs, the intended payee sends a message to the payor that is akin to an invoice for payment, together with evidence that the obligation was met. Being the blockchain, these messages are public prior to being committed to a block. Also, as there is potential congestion on the network, a message is sent with a delay depending upon the transaction fee nominated by the payee. In the intervening time, front-runners, or bots programmed to front-run, see the message and can resend it, substituting their own address for payment and a higher transaction fee to achieve priority (Daian et al., 2019; Eskandari et al., 2019). The payor's account for that contract is then drained of tokens before the intended payee can be paid.[3]

While such front-running is akin to the leap-frogging activities of high-frequency traders,[4] in this case, it threatens to undermine the ability to offer smart contracts on any blockchain system.[5] Fearing non-payment, a contract payee, may not perform or enter into a contract at all. This harms both parties and will likely stifle the development of smart contracts and the ensuing gains from trade. While some solutions involving encrypting messages have been posited, these can only potentially assist in some bilateral contracts between known and identifiable parties (Copeland, 2021) unless implemented at a platform level. Other solutions involving increasing the transparency of 'front-running' races do not actually resolve the problem and merely place the payee on a more level playing field than front-running bots.[6]

---

[3] The total value of tokens gained in this manner is estimated at almost $1 billion since January 2020 (https://explore.flashbots.net/) although the vast majority of that is from arbitrage front-running rather than liquidation front-running which is the focus of this paper. See also, Ferreira Torres et al. (2021).

[4] This occurs where a large trade is placed, and bots can trade in the market ahead of that trade and exploit arbitrage opportunities. This happens for cryptocurrencies on the blockchain as well, using a technique called 'insertion' to front-run high-value transactions; Ferreira Torres et al. (2021). However, this is not the type of front-running considered in this paper.

[5] See Robinson and Konstantopoulos (2020). The problem was first identified in 2014 in a Reddit post from pmcgoohan; see Stankovic (2021) for an overview. It is also possible that this activity could undermine the consensus layer of blockchains through front-running on past blocks using a time bandit attack; Daian et al. (2019).

[6] https://ethresear.ch/t/flashbots-frontrunning-the-mev-crisis/8251 and the critique by Ed Felten, https://medium.com/offchainlabs/meva-what-is-it-good-for-de8a96 c0e67c. Such auctions may also reduce the congestion effects generated by front-running; Buterin response.

6.2   A Model of Front-Running

As a first step, it is important to understand the mechanics of front-running and the incentives of those engaging in this behaviour. The unit of analysis is a given contract. That contract comprises certain performance obligations whose performance can be verified by party a sending a message $M_a = \{\alpha, E\}$ where α is a's wallet address and E is verifiable evidence of performance to the network as a transaction. That transaction is then confirmed to a block and recorded on the blockchain. At that point, any payment, T, triggered by the receipt of M_a involves T in tokens being transferred to a's wallet. Note that any agent, i, sending a message that is confirmed to a block specifies and pays a transaction fee, $f_i > 0$.

Front-running arises when b observes M_a as it is broadcast to the network but before it is confirmed to a block, and b chooses to send a message $M_b = \{\beta, E\}$ to the network. If M_b is confirmed to a block ahead of M_a, then T is automatically sent to b's wallet, and a receives no payment. This could arise if M_b is confirmed to a block earlier than the block M_a is confirmed on or if it is confirmed to the same block with an earlier order amongst transactions in that block.

Given that M_a is broadcast first, how could M_b be recorded on the blockchain with an earlier time-stamp? Note, first, the messages are initially broadcast to a mempool. Those transactions are then validated and confirmed by miners or validating nodes who are responsible for ordering the transactions.[7] All valid transactions are recorded on the blockchain, at which point the transaction with the earliest time-stamp will trigger the contracted actions. Miners or validating nodes will then choose the order of transactions. On the Ethereum blockchain, miners will try and maximise transaction fee revenue by prioritising transaction recording based on the transaction fee bids (or offered 'gas') that accompany a message. Thus, M_b can, by offering to pay a higher transaction fee, be ordered ahead of M_a in a block. Of note is the fact that, because miners have the power to order transactions, to the extent that transaction ordering matters, the ability to earn payments based on ordering power

---

[7] Miners are responsible in Proof of Work protocols while validating nodes are responsible in Proof of Stake protocols.

has been termed the *miner-extractable value* (or MEV) (see Daian et al., 2019).[8]

It is useful to illustrate the severity of this issue for contracting. Suppose that a party contracts a to provide a service using a contract recorded on the blockchain. If a performs the contract, assume that it costs them, $c > 0$, to do so, and T will be paid if evidence of performance is submitted. In the absence of front-running, a sends a message, M_a on the blockchain for a fee f_a that can be arbitrarily small and ends up with a payoff of $T - c - f_a$ which is assumed to be positive.

Suppose now that front-running can occur. If b sends a message M_b for a fee of f_b they can potentially earn T. If $f_a = f_b$, then b's expected payoff is $\frac{1}{2}T - f_b$ and a's falls to $\frac{1}{2}T - c - f_a$. In effect, the payment to a is taxed at 50 per cent assuming only one front-runner. If there is more than one, the effective tax is higher.

This analysis, however, assumes that transaction fees are fixed. However, these are chosen by agents recording transactions on the blockchain. Typically, if $f_b > f_a$, b would be recorded at an earlier time-stamp and their payoff becomes $T - f_b$ while a's drops to $-c - f_a$. In reality, a and b choose their fees as part of a first-price, sealed bid, all-pay auction for priority. Here, b will choose a fee up to $f_b = T$ requiring a to exceed that to achieve priority, something that is not worthwhile. Having both post fees equal to T is not an equilibrium outcome if priority is then randomly assigned. Instead, one sets a fee at T while the other sets an arbitrarily low fee or does not choose to send a message. Given this, either a sets $f_a = T$ and earns a payoff of $-c$ or it sets a low or zero fee and earns the same payoff. Under these conditions, a chooses not to incur c and perform the contract regardless of how high T is.

Under these conditions, contracts that require settlement on the blockchain will not arise in equilibrium. Various solutions have been proposed to mitigate such issues. These have included auctions to make priority a more transparent process (Buterin (2021; Daian et al., 2019). However, these auctions do not prevent front-runners from participating, and that competition still immiserises contract safety as outlined above. A second solution involves adjusting blockchain protocols to improve time-stamping. However, as there are always lags of some kind achieving this is difficult. A third set of solutions involves encrypting messages until they

---

[8] For a demonstration of a smart contracting being front-run in this manner see Scott Bigelow, "How To Get Front-Run on Ethereum mainnet", YouTube June 17, 2020.

are confirmed on the blockchain (Aune et al., 2018). This can resolve this problem, but it requires implementation at the blockchain protocol level, encrypting all messages, which is computationally expensive and moves away from the public nature of blockchain interactions.

Compared with contracting outside of the blockchain, the reason why such front-running is a threat is that there is no proof of identity required for payment. This is by design, as the privacy of parties on the blockchain is a feature. Thus, blockchain smart contract systems are characterised by contracts that specify performance objectives but not the identity of those performing them. This allows anonymity in payments to be preserved. If the contract specified that following performance, payments would be made to a's wallet (α) specifically, front-running could not occur as M_b would not trigger a payment to β. However, anonymity means that the addressee for payment can be substituted without altering the contract. Our examination here is made on the basis that this blockchain feature needs to be preserved.

The following assumption is made about agent preferences: the agent who has actually performed the contractual obligation and broadcasts a message of that performance earns T in utility if they receive payment for that performance, $-\theta$ if that payment goes to an illegitimate claimant and 0 otherwise.

6.3   USING CRYPTOGRAPHY

Front-running arises because transactions are visible in the mempool before they are confirmed to a block. Cryptography can be used to encrypt the transaction content until they are confirmed on a block at which point it can be decrypted and read. Thus, no transaction is visible to front-runners.

The decrypting process can be done by an individual smart contract if the parties have an off-chain relationship and can be identified by their public keys. It can also be conducted by validating nodes who coordinate off-chain regarding the decryption keys and decrypt transactions if a certain share of them (say, two-thirds) provide the requisite keys.[9]

There are two main issues with using cryptography in this way. First of all, for network solutions, it requires encryption of all transactions

---

[9] See, for instance, the discussion in Malkhi and Szalachowski (2022).

and messages, which creates additional, non-trivial computational costs. Second, these solutions require off-chain coordination of some description which moves away from the fully automated objectives in some blockchains. Nonetheless, given these constraints, cryptography can solve the front-running problem.

6.4   A Mechanism to Deter Front-Running

Having set up the front-running problem, is it possible to design a mechanism to deal with the issue? The first step is to understand that front-running of the type considered above is an example of an ownership dispute. It falls into a class of ownership disputes exemplified by the biblical dispute heard by King Solomon over the 'ownership' of a baby (Gans & Holden, 2022). Such disputes have the following characteristics:

1. The legitimate claimant is part of the set of agents making an ownership claim.
2. Legitimate and illegitimate claimants know if they are legitimate or not.
3. Legitimate and illegitimate claimants have different preferences over who, other than themselves, are allocated ownership.

In the case of Solomon's adjudication over who was the true mother of a baby, it was known that the true mother was one of the two claimants; each claimant knew their own status and, as we will discuss, it was a feature of the story that the true mother had different preferences than the other agent over what happened to the baby if ownership was not allocated to them. For blockchain front-running, the nature of the problem necessitates the legitimate claimant being part of the relevant claimant set, claimants knowing their own status, and illegitimate claimants being indifferent to other outcomes which may not be the case for the legitimate claimant.[10]

---

[10] The mechanism described here is a simple, special case of the Simultaneous Report (SR) mechanism developed by Chen et al. (2022) that itself is a simplification of mechanisms explored by Moore and Repullo (1988) and Moore (1992). It is, however, distinct from the divided ownership processes examined by Ayres and Talley (1994) as it envisages a solution outcome whereby ownership is not divided.

The Need to Discretise Time

In the literature on front-running in financial markets, one proposed solution was to change the time on an exchange from continuous to discrete time (Budish et al., 2015). To operate a mechanism involving multiple agents to resolve front-running on blockchains, we must similarly discretise time. This is done by the smart contract proposing a time period counted from the time a first message M_i is recorded on the blockchain during which all such messages are collected, and the mechanism we propose is run on them. The length of the time period, let's call it Δ, is a parameter that can be chosen.[11] Increasing Δ reduces the need for claimants to pay higher transaction fees so as to participate in the mechanism but also results in a delay in payment. If there is a single message received during Δ, there is no ownership dispute over the payment, and the payment is made to the wallet addressee. If there is more than one message received, there is a dispute, and the mechanism is run to resolve the dispute immediately upon Δ's end.

The Single Legitimate Claimant Case

While our mechanism applies for an arbitrary number of ownership claimants to T, initially, it is assumed that there are two claimants, a (the legitimate claimant) and b (the would-be front-runner or illegitimate claimant). Each claimant knows their own status, but this is known to the mechanism designer. The designer's goal is that the payment only is made to the true claimant.

The process is initiated once a claim is validated and confirmed on the blockchain. Consider the following mechanism:

1. If, in a time period, Δ, there is a single message, $\{i, E\}$, send T to i.
2. If, in a time period, Δ, there are two messages, $M_a = \{\alpha, E\}$ and $M_b = \{\beta, E\}$, the challenge stage begins (Fig. 6.1 illustrates the process by which claims are assembled on a blockchain).

---

[11] Δ can be measured in time units or in blocks, with the first block being the one where a claim(s) is first confirmed.

The **challenge stage** involves:

1. One agent is chosen at random and given the opportunity to withdraw their claim.
2. If the claim is withdrawn, the other agent is paid T.
3. If the claim is asserted, T is paid to a third party (or, equivalently, the tokens burned), and the contract is nullified.

Thus, as is depicted in Fig. 6.1, the legitimate claimant first broadcasts a message to the mempool where it can be seen by others triggering illegitimate claims. All claims pay the requisite fees and are confirmed to blocks in the specified time period, Δ. The mechanism is then run drawing from confirmed claims.

Given this, we can prove the following:

Proposition 5 Suppose that $\theta > 0$. The unique subgame perfect equilibrium involves a single claimant who is the agent who performs the obligation.

Proof Without loss in generality, suppose that a is the true claimant, and the challenge state is initiated as b also makes a claim. Thus, both agents have incurred the transaction fee, f. There are two cases to consider:

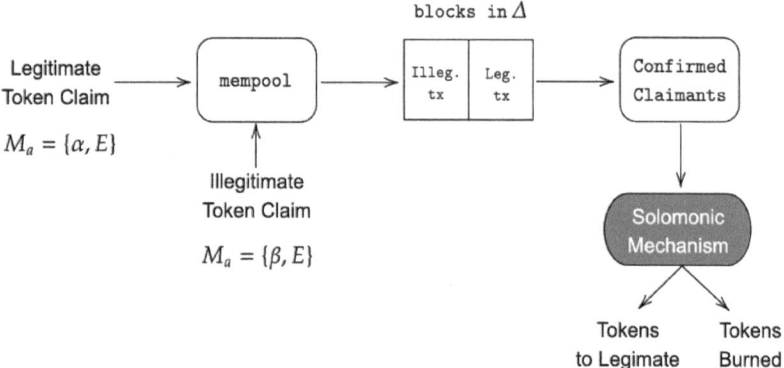

Fig. 6.1 Assembling competing claims

1. If a has the opportunity to renounce their claim, they will receive $-\theta$ if they renounce their claim (as they know the other claimant is illegitimate) and 0 otherwise. Thus, if $\theta > 0$, a will continue to assert their claim, and T will be sent to a third party, with each agent ultimately earning $-f$.
2. If b has the opportunity to renounce their claim, they will receive 0 regardless (as they know the other agent is legitimate). Thus, they will be indifferent between renouncing or not, and their ultimate payoff will be $-f$.

We now examine each agent's incentive to make a claim. As a moves prior to b, we work backwards by considering b choice. If b makes a claim (by front-running), they expect to earn $-f$ as a will never renounce their claim if $\theta > 0$. Thus, b will not make a claim. Given that b will not make a claim, a will make a claim and earn $T - f$.

Note that the mechanism requires that the true claimant have a strict preference regarding whether a payment is made to an illegitimate claimant. Otherwise, if there is a possibility that the legitimate claimant may renounce their claim, this opens up a potential return to illegitimate claimants. A weaker assumption that leads to this same outcome would be that if a true claimant is indifferent as to where the payment is made, if not to themselves, they choose to assert the claim. In equilibrium, if the true claimant were programming in their assertion response at the time they submit M_a, then it is optimal for them to assert their claim. Thus, the $\theta > 0$ assumption does not play a role if agents pre-commit their mechanism responses as might arise in a Blockchain environment.[12]

It should be emphasised that the Solomonic mechanism proposed here is just one type of multi-stage reporting mechanism that might be used to deter front-running. For instance, Canidio and Danos (2022) propose a 'commit-reveal' approach. This involves an initial message sent by the legitimate claimant that involves a hash of both their address and the evidence for the smart contract. (Recall that such hashing makes it impossible for others to observe the evidence in the mempool, and the inclusion

---

[12] Since Aghion et al. (2012) it has been understood that certain mechanisms may not be robust to small perturbations from common knowledge. Chen et al. (2022) shows how suitably designed lotteries can ensure that the mechanism used in this paper is robust to private-value perturbations. It would be straightforward to do so in this environment if desired, although it would make the mechanism slightly more involved.

of the address makes the hash itself not useful to them). The smart contract logs this message, and then the legitimate claimant sends the public message with their address and evidence. This message could be subverted by front-runners. However, Canidio and Danos (2022)'s mechanism requires the smart contract to check any claim with the previous hashed message. Only the legitimate claimant has the correct sender address, and so when the smart contract hashes any new message and compares it with the previous one, only the legitimate claimant will be identified as the correct agent to send any payment. This approach is not dissimilar to the two stages of the Solomonic mechanism, although in that case, in equilibrium, only one message is sent. By contrast, the commit-reveal mechanism involves two messages always. This illustrates the trade-off in using cryptographic tools (like hashing) versus using incentives again. The former bakes in any costs while the latter can, if an equilibrium is established, avoid them. This also means that, depending on the smart contracting environment, one approach may likely be superior to another.

Further Issues

The mechanism also yields a single true claimant if there are many potential illegitimate claimants. The only difference is that one claimant out of the pool of claimants is allowed to renounce their claim in the challenge stage. Because the true claimant knows they are the true claimant, they will also choose to assert their claim if $\theta > 0$.[13]

There is a possibility that there could be multiple claimants, with the illegitimate claimants forming a coalition. In this case, a randomly selected agent could choose to withdraw their claim, but there still be multiple claimants. In this case, what would happen to the tokens? One way of overcoming this is when there are more than two claimants, a set of agents is randomly selected. If any assert their claims, T is paid to a third party. If all withdraw their claims, those agents are removed from the pool of claimants, and a new set of agents (half or just under one-half of the

---

[13] One possible front-running strategy would be for a front-runner to send multiple messages for the same wallet address. To avoid this, a claim to be resolved would draw based on messages. Of course, front-runners could send messages to different wallets. This, however, would not exclude the true claimant and so would ultimately fail.

remainder) are randomly selected, and the mechanism is repeated. Eventually, an agent who is asserting their claim will be selected, and T will be paid to a third party. There is guaranteed to be one such agent as the true claimant is amongst the starting pool.[14]

What if, due to network issues, the true claimant is not amongst the pool of claimants when the mechanism begins? If illegitimate claimants believe that this is a possibility, they have an incentive to make such claims. Clearly, if they are the only claimant, they will be able to capture T. If they are amongst multiple claimants, this is not possible unless those multiple claimants are controlled by them. Thus, there would have to be some collusive mechanism amongst illegitimate claimants to subvert the mechanism in this way. So long as the probability that the legitimate claimant is not in the relevant pool at the time the mechanism is run is low enough, the deterrence effect of the mechanism remains.

An even simpler version of this mechanism relies on 'mutually assured destruction' by automatically burning the tokens if there is more than one claimant.[15] A practical feature of the Solomonic mechanism is that an illegitimate claimant must pay an additional transaction fee to send the message at the challenge stage, as this must be written to the blockchain. Such a fee also breaks their indifference, while the true claimant has preferences that make them willing to pay a certain fee to assert at the challenge stage and ensure that an illegitimate claimant does not receive the tokens.

The Multiple Legitimate Claimant Case

The above analysis envisages a contract on the blockchain where there is only one agent who can be the legitimate claimant. However, in some applications—say involving bounties or rewards for performance—can have multiple legitimate claimants. In the absence of front-runners, such rewards would be made based on some time verification of the messages from agents. That may result in multiple claimants, but the contract could specify a tie-breaking rule or another measure to award

---

[14] If there were concerns that this process may take time, then a cost could be imposed on each round of participation.

[15] This is a reasonable approach, but our challenge stage permits the construction of lotteries mentioned in the previous footnote that makes it the *unique* optimal choice for an illegitimate claimant to withdraw their claim (rather than being indifferent).

the bounty, including splitting it. When there are illegitimate claimants, however, those rules would create an incentive to front-run the contract.

A potential solution, in this case, would be to run the mechanism as proposed for the one legitimate claimant case. This might be done by shortening the time (Δ) where claims will be evaluated even if this results in potentially higher transaction fees. Reducing Δ means that any true claimant will be more likely to believe that no other legitimate claimant will submit a competing claim during that period and there will be one true claimant. In that case, the fact that front-runners do not have an incentive to claim will preserve the contractual incentives. The cost is that this will limit the tie-breaking options that might otherwise be used in such contracts. Such options are important if they play a role in providing incentives to compete and perform the contract obligations.

There is, however, a counter-risk that arises in this particular case: the payor may have an incentive to front-run their own mechanism. This would arise if it could not be guaranteed that T was being sent to a party other than the payor. Thus, the mechanism would have to specify that the tokens be burnt or sent to a legitimate charity account that is publicly verified.

These potential issues, however, need to be weighed against the real possibility that the contract would be otherwise completely unworkable if front-running was possible. Thus, the use of the mechanism expands the feasible contract space but does not obtain the full range of options that would be available if front-running were not possible at all.

Implementation Choices

The mechanism we analyse here can be easily implemented on existing blockchains. Indeed, we have already provided code for a generic smart contract on the Ethereum network.[16] In effect, it is a Solomonic clause added to existing smart contracts.

---

[16] See the repository at https://github.com/solomonic-mechanism.

There are many open design choices in implementing Solomonic clauses that we list here but that their resolution is beyond the scope of the present paper. These include:

- *Hard-coded challenge response*: the mechanism as outlined includes a claim message followed by a message in the challenge stage if a potential claimant is selected. However, it could be envisaged that the initial message contains the contingent response in the challenge stage rather than requiring a separate message and fee payment.
- *Randomisation*: implementing randomisation on a blockchain virtual machine can be challenging and often requires a call to an Oracle that is off-chain. In our implementation example, the agent chosen in the challenge stage was not chosen at random but was the agent with the most recent time-stamp recorded on the blockchain. Theoretically, this is the agent most likely to be the true claimant, as they would not be putting forward higher transaction fees as part of a front-running strategy. However, due to latency on the internet, that agent is, in part, determined randomly and thus, this would be a useful alternative to pure randomisation.
- *Time period*: in our implementation, we set the time period, Δ, to 60 seconds (or four blocks on Ethereum). The actual time period chosen would depend on other factors, including network congestion and the need to clear token payments quickly or not.
- *Token burning*: If a claim is asserted in a challenge stage, then the tokens need to be transferred away from any party in the arrangement for the mechanism to work. This could involve burning the tokens (sending them to a null address) or, alternatively, having the tokens become part of a fund or non-profit. As the mechanism, if successful, should involve little of this in equilibrium, where the tokens are sent is a decision that should be made to ensure that the mechanism is not attacked by malicious agents trying to force the tokens to be burnt or otherwise undermine the operation of the smart contract.
- *Signalling*: A contract with a Solomonic clause could involve a message for payment indistinguishable from contracts without that clause or one that indicated the existence of the Solomonic clause. The distinction would impact front-running and its attempts. When there is a clear signal, front-runners will avoid these contracts. When

there is no clear signal, they may not unless there is a sufficient share of contracts with a Solomonic clause, in which case, front-running on all contracts may not be worthwhile. The use of such signals is, therefore, an important implementation choice.

6.5   CONCLUSION

This chapter has illustrated two broad approaches when parties to a transaction make public disclosures that can be used as the basis for an attack. Cryptography hits the issue at its source by hiding information from view and denying attackers the opportunity to exploit it. An incentive mechanism takes an indirect approach that reduces the payoff from an attack to make it not worthwhile for attackers to disrupt transactions. The former approach, however, is costly, while the latter approach can take some time to establish it as an equilibrium.

The case of front-running on blockchains illustrates the broad trade-off. However, there likely exist incentive mechanisms that can act as a substitute for cryptography to make interactions safer. Those mechanisms have not received much attention in the computer science literature on blockchains.

Key insights from Chapter 6

- Blockchain front-running can make it impossible to operate smart contracts by making it difficult for parties that perform contractual obligations to receive payment for their actions. Instead, that payment is hijacked by others observing the relevant requests in the mempool and bidding for priority in confirming those transactions to blocks.
- Cryptography can eliminate such front-running but encrypting transactions on the mempool until they are confirmed in blocks. However, this raises the (computational) cost of transactions.
- A Solomonic mechanism that automatically handles disputes over payments can deter attackers while making transactions safe for legitimate contract parties. However, these must be established as an equilibrium in order to make transactions safe from attack.

References

Aghion, P., Fudenberg, D., Holden, R., Kunimoto, T., & Tercieux, O. (2012). Subgame-perfect implementation under information perturbations. *The Quarterly Journal of Economics, 127*(4), 1843–1881.

Aune, R. T., Krellenstein, A., O'Hara, M., & Slama, O. (2018). Footprints on a blockchain: Trading and information leakage in distributed ledgers. *The Journal of Trading, 13*(4), 49–57.

Ayres, I., & Talley, E. (1994). Solomonic bargaining: Dividing a legal entitlement to facilitate coasean trade. *Yale LJ, 104,* 1027.

Budish, E., Cramton, P., & Shim, J. (2015). The high-frequency trading arms race: Frequent batch auctions as a market design response. *The Quarterly Journal of Economics, 130*(4), 1547–1621.

Buterin, V. (2021, June). Proposer/block builder separation-friendly fee market designs.

Canidio, A., & Danos, V. (2022). *Commitment against front running attacks.* Technical report.

Catalini, C., & Gans, J. S. (2020). Some simple economics of the blockchain. *Communications of the ACM, 63*(7), 80–90.

Chen, Y.-C., Holden, R., Kunimoto, T., Sun, Y., & Wilkening, T. (2022). Getting dynamic implementation to work. *Journal of Political Economy, 131*(2), 285–387.

Copeland, T. (2021, November). A dex on cosmos is working on a way to prevent front running. *The Block.*

Daian, P., Goldfeder, S., Kell, T., Li, Y., Zhao, X., Bentov, I., Breidenbach, L., & Juels, A. (2019). *Flash boys 2.0: Frontrunning, transaction reordering, and consensus instability in decentralized exchanges.* arXiv preprint arXiv:1904.05234

Eskandari, S., Moosavi, M., & Clark, J. (2019). *SoK: Transparent dishonesty: Front-running attacks on blockchain.* Springer.

Ferreira Torres, C., Camino, R., et al. (2021, August 11–13). Frontrunner jones and the raiders of the dark forest: An empirical study of frontrunning on the ethereum blockchain. In *USENIX Security Symposium, Virtual.*

Gans, J. S. (2022). The fine print in smart contracts. In M. Corrales Compagnucci, M. Fenwick, & S. Wrbka (Eds.), *Smart contracts technological, business and legal perspectives* (Chapter 2). Hart Publishing.

Gans, J. S., & Holden, R. (2022). *A solomonic solution to ownership disputes: Theory and applications.* Technical report, Mimeo.

Gans, J. S., & Holden, R. (2023). A solomonic solution to blockchain frontrunning. *American Economic Association: Papers and Proceedings, 113,* 248–252.

Holden, R., & Malani, A. (2021). *Can blockchain solve the hold-up problem in contracts?* Cambridge University Press.

Malkhi, D., & Szalachowski, P. (2022). *Maximal extractable value (MEV) protection on a DAG.* arXiv preprint arXiv:2208.00940

Moore, J., & Repullo, R. (1988). Subgame perfect implementation. *Econometrica, 56*(5), 1191–1220.

Moore, J. (1992). Implementation, contracts, and renegotiation in environments with complete information. *Advances in Economic Theory, 1,* 182–282.

Robinson, D., & Konstantopoulos, G. (2020, August). Ethereum is a dark forest. *Medium.*

Stankovic, S. (2021, August). What is MEV? Ethereum's invisible tax explained. *Cryptobriefing.*

CHAPTER 7

Rules Versus Mechanisms

Abstract This chapter explores the potential of mechanisms as an alternative to rules in achieving consensus on blockchains. While rules offer simplicity, they can limit flexibility and fine-tuning. Mechanisms, on the other hand, provide greater adaptability and can enhance efficiency and sustainability. By integrating computer science and economic approaches, mechanism-based consensus protocols can offer advantages over existing Proof of Stake protocols, allowing users to manipulate trade-offs between finality and liveness. By examining the potential applications of mechanisms in both BFT and LCR-based Proof of Stake networks, the chapter highlights the expanded set of blockchain consensus options provided by these approaches.

Keywords Rules · Mechanisms · Blockchain consensus · Proof of Stake · Mechanism design

All of the current protocols to achieve blockchain consensus rely on rules. Of course, those rules also define certain games—particularly coordination games—that participants in a network play. Those games will determine whether networks can be manipulated and ultimately will operate as intended and be a reliable source of data to enable many economic and non-economic activities.

© The Author(s), under exclusive license to Springer Nature
Switzerland AG 2023
J. Gans, *The Economics of Blockchain Consensus*,
https://doi.org/10.1007/978-3-031-33083-4_7

However, faced with situations where there is the possibility of strategic manipulation, economists sometimes propose not relying on rules to contain behaviour but instead propose mechanisms to induce better behaviour. Rules have the advantage that they can be easily encoded but have the disadvantage in that they make it hard to achieve fine-tuning at the margin. Thus, the trade-offs examined thus far in the book become starker than they need to be.

Mechanisms, while more complex than rules, can potentially allow for that fine-tuning. This chapter examines their potential and provides a background to the type of mechanisms that could be implemented in Proof of Stake blockchains—both permissioned and permissionless—to achieve more efficient outcomes and create more sustainability. In this respect, there is a potential future where computer science and economic approaches are far more integrated and complementary in generating consensus on blockchains.[1]

7.1   What Is Blockchain Truth?

As we have seen, blockchains rely on transaction messages being broadcast (to the mempool) where agents (validating nodes) assemble them into a block that is appended to a public ledger. Those agents confirm that messages are valid (e.g., tokens transferred from a wallet actually are owned initially by that wallet) and the block is accepted as confirmed by other agents (i.e., that there is consensus regarding the validity of the proposed block). There are two challenges: (i) that consensus is reached and (ii) that consensus is over a set of truthful messages. For distributed ledgers, 'truth' has a specific meaning: that the messages assembled into the block are those in the mempool without any being privately added by validators or intentionally excluded (or censored) by them.

In previous chapters, we have examined both BFT and LCR methods of achieving consensus. While these methods achieve consensus, it is safe to say that consensus on the truth is left to be determined by crowd behaviour. For instance, in LCR blockchains, there can be forks where two equally long chains exist with the potential that at least one of them has untruthful blocks—e.g., past blocks that exclude messages in order to facilitate a double-spending attack. In BFT blockchains, attacks

---

[1] This chapter is based on and draws liberally from Gans and Holden (2022).

that subvert the operation of the blockchain by allowing consensus to be delayed are possible. In each case, so long as the majority or super-majority of participants (weighted by computational power in Proof or Work or token holdings in Proof of Stake) are engaging in truthful messaging—that is, messages that reflect what has been broadcast to the mempool—then truthful consensus can be achieved in equilibrium. Nonetheless, in trying to achieve coordination in this way, there is the potential for some disruption if adverse agents coordinate interventions.

Are there more efficient and reliable ways to achieve truth in consensus by designing and encoding mechanisms? Mechanism design is the branch of economics that deals with creating incentives for self-interested agents with information not known to the designer to reveal that information truthfully and still be willing to participate in the relevant economic activity. For example, auctions can be viewed as mechanisms for the truthful revelation of the buyers' willingnesses to pay to a seller. Without a mechanism, buyers will not want to reveal their true value to a seller who might take advantage of that by charging a higher price. Similarly, without a mechanism, a seller has to guess buyer valuation when setting a price lest buyers choose not to buy or delay purchases. An auction, by specifying how a bid (a reflection of a buyer's true valuation) can be used, can be designed in such a way that the buyer has an incentive, to tell the truth. This benefits buyers and the seller compared to a mechanism-less alternative but, critically, it relies on the mechanism being followed by the seller (Akbarpour & Li, 2020) and bid information being communicated accurately (Eliaz, 2002; Hurwicz, 1972).

Typically, mechanisms are conceived of as being designed and then implemented in a centralised manner. This would, on the face of it, put it at odds with being used in permissionless blockchains whose modus operandi is decentralised operation. However, while permissionless blockchains are often characterised as decentralised—i.e., no one entity controls them, and there is no one single point of failure—a key part is centralised by virtue of there needing to be consensus on the state of the blockchain. Moreover, the blockchain's code is public, and itself regarded as immutable. Thus, the ingredients for both a theoretical (i.e., to obtain truthful revelation using incentives) and practical use (i.e., transparent, unchangeable and unique code) are present for the use of mechanisms. This motivates our current examination of that possibility.

There is potential to use mechanisms in Proof of Stake blockchains. In Proof of Work, the scope to deploy a mechanism does not exist because winning the computational game gives the proposer an unfettered right as to the block proposed. In Proof of Stake, by contrast, validating nodes are required to have committed a number of tokens prior to potentially being selected to propose a node. That stake (technically, a bond) can then be used to provide incentives for any mechanism—for instance, creating a risk that if the node does not propose a truthful block, that node will be worse off than had not participated in the mechanism at all.

Indeed, for behaviours that can be readily identified as illegitimate—such as proposing two conflicting blocks, being unavailable after promising to be available or proposing a block that isn't chained to the genesis block (or a checkpoint)—as was discussed in Chapter 5 Proof of Stake blockchains automatically fine misbehaving nodes using slashing (Buterin, 2017). This relies on the protocol designers identifying specific behaviours that may be associated with ill intent rather than something more straightforward and robust. Here the goal is to examine whether messages exist to ensure that validating nodes propose truthful blocks more generally using the information that exists amongst all nodes.

This chapter provides two mechanisms that can be used for BFT and LCR blockchains respectively to potentially ensure that consensus outcomes are also truthful. The purpose here is to show the potential benefits of mechanisms to aid in achieving truthful consensus on blockchains. Of necessity, this requires considering simplified environments and, thus, abstracting away from practical difficulties associated with coding such mechanisms. However, this broad framework could be used as the foundation for the practical implementation of mechanisms on blockchains and improve their operation.[2]

7.2   Mechanism for Byzantine Fault Tolerance

The first broad consensus mechanism is Proof of Stake under BFT. In the absence of an economic mechanism achieving consensus, under BFT we have seen that a voting mechanism is used. Recall that these voting mechanisms have the following steps:

1. Transactions are broadcast as messages to the mempool
2. Nodes stake and commit to be part of the validating pool

---

[2] This chapter draws liberally from Gans and Holden (2022).

3. Nodes observe messages
4. One node is selected to propose a block
5. Other nodes choose whether to confirm or reject the proposed block
6. If at least two-thirds of nodes confirm the block it is accepted otherwise it is rejected, and another node is selected to propose a block, and the process begins again

Typically, in voting to confirm a block, nodes check the technical validity of the proposed block and also whether other nodes are confirming the same block. Thus, communication is multi-lateral and network-wide in the process of achieving consensus. Here we consider whether a mechanism can replace the voting process and limit communication to just two randomly chosen nodes before appending a new block to the chain.

A Simultaneous Report Mechanism

The mechanism examined here is a special case of the Simultaneous Report (SR) Mechanism analysed by Chen et al. (2022).[3] The baseline idea is that messages are broadcast publicly by blockchain users to the network and participating nodes assemble them into blocks of a fixed size based on the time broadcast. When a block is proposed to be committed to the blockchain, each node has in its possession a block of messages they have received. It is assumed that this block is common across all nodes, however, there are no restrictions on nodes in proposing an alternative block. The goal is to ensure that nodes, while able to propose alternative blocks, only propose and accept truthful blocks.

Suppose there are nodes, $i \in \{1, ..., n - 1\}$ each of whom assemble ledger entries into a block of fixed size. If a node ends up proposing a block that is accepted, they receive a block reward, R. There is also an nth node that proposes a manipulated block. If that block is accepted, they receive a payoff θ that is private information in addition to the block reward, R. Nodes can send any message from a countably infinite set.

Consider the following mechanism that is run after messages have been sent to the mempool:

---

[3] The SR mechanism is a simplification of the multi-stage mechanisms explored by Moore and Repullo (1988).

1. One node is randomly chosen to be the *proposer*, p, and another node, c, is chosen to be the confirmer.
2. The proposer proposes a block in the form of a message, M_p while the confirmer sends a message, M_c.
3. If $M_p = M_c$, then the block is committed to and added to the blockchain. The proposer receives R.
4. If $M_p \neq M_c$, then the challenge stage begins with both p and c being fined, $F > 0$.

The **challenge stage** involves:

1. p sends a new message M_p^C based on knowledge that there is a disagreement.
2. If $M_p^C = M_c$, then M_p^C is committed to the blockchain, p receives R, and c is refunded F.
3. If $M_p^C \neq M_c$, then p's proposal is discarded and the process begins again with p and c excluded from subsequent rounds.

Given this, we can prove the following:

Proposition 6 Suppose the true block is M_T. Then the unique subgame perfect equilibrium outcome for the mechanism for any pair of nodes is $M_p = M_c = M_T$.

Proof Suppose that the selected pair does not include node n. Then working backwards, if $M_p \neq M_c$, then $M_p \neq M_T$, $M_c \neq M_T$ or both as the message space is a (countably) infinite set. In this case, the challenge stage is initiated and p has the opportunity to send a new message. If $M_p = M_T$, then there is zero probability that p could send $M_p^C = M_c$ and so p's proposal is discarded and both nodes receive $-F$. If $M_p \neq M_T$, then by selecting $M_p^C = M_T$, then with some probability (possibly equal to 1), p receives $R - F$ rather than $-F$ with certainty by choosing some other message. Thus, in the challenge stage $M_p^C = M_T$. Anticipating this, it is optimal for p to set $M_p = M_T$ and c to set $M_c = M_T$.

Now suppose that the selected pair includes node n. If node n is the confirmer and the challenge stage is reached, then, we have already shown that the proposer will set $M_p^C = M_T$. Given this, node n will find it optimal to set $M_c = M_T$ and earn 0 rather than $-F$.

Alternatively, if node n is the proposer, by our earlier argument, the other node will set $M_c = M_T$. If n sets $M_p \neq M_T$, then there is a challenge round. In that round, n will earn $R - F$ by setting $M_p^C = M_T$ and $-F$ otherwise. Given this, it is optimal for n to set $M_p = M_T$ as it will earn R rather than $R - F$.

It is easy to see that in the challenge stage if M_c is the truth, p knows this and so finds it worthwhile to set $M_p^C = M_c$ and receive R. If M_p is the truth, p has a problem as it does not know what M_c was. In this case, it ends up setting $M_p^C \neq M_c$ and receiving 0. Thus, the truth is revealed regardless. There seems something harsh about this last step as the proposer may be reasonable and still punished. That would arise only if c has an incentive to message something other than the truth. If they message the truth, then they get 0 as they expect p to revise their message at least and receive a refund of F. If they message something else, then p will never be able to guess that and so they will lose F. Thus, c has no incentive to do anything other than be straightforward.

It is useful to stress how remarkably powerful this mechanism is for obtaining consensus on truthful blocks. There are several features that should be noted. First, F can be arbitrarily small and the true block will be confirmed by any pair. Second, any randomly selected pair is sufficient to confirm a block. Unlike BFT mechanisms, there is no need for multiple confirmation rounds, pre-commits or messages sent from more than two nodes. Once block transactions have been communicated publicly and formed into the truthful block, the mechanism can take place, and confirmation is instantaneous. Third, if there is more than one node that has a private value that arises should a block other than the truthful block occur, the true block will still be confirmed. This outcome occurs even if two nodes with private preferences happen to be paired. This will happen so long as those nodes have preferences for distinct blocks. However, as we explore in the next section, if nodes have preferences for the same non-true block, the game is more complex (e.g., if there is a coalition of nodes). Third, a key part of the mechanism is that while all nodes have common knowledge of the message for the true block, privately preferred blocks are unknown beyond individual nodes. This subverts any method by which coordination could arise on a non-true block. Once again, a coalition of nodes with a non-true preferred block could potentially coordinate and subvert the mechanism.[4] Finally, it is useful to note some

---

[4] It is worth emphasising that the mechanism does rely critically on the messages of the true block being perfect and common knowledge. If this was not the case, the proof

practicalities in terms of implementing this mechanism. The mechanism relies on the two nodes being selected randomly. As will be discussed in detail below, randomness plays an important role in the mechanism working when there is more than one attacking node. Finding a pure randomisation device on-chain is a challenge for blockchains and, thus, it is likely that this part of the mechanism will rely on an external randomisation input. The mechanism itself can run on-chain but it would be as a smart contract coded into the protocol. This is something that is a feature of many Proof of Stake protocols for other elements.

Robustness to Multi-node Attacks

As noted above, while the proposer, if selected, has an incentive, to tell the truth, this is based on a specific assumption that could not be guaranteed for a permissionless blockchain (and maybe not all permissioned ones either): that the proposer and confirmer are different entities. What if the proposer and confirmer are the same person or part of an attacking coalition that share the same incentive to confirm an alternative, non-true block?

Suppose that an attacking coalition has a share, s, of all nodes. If they are the proposer, then, with probability s they will be able to confirm the distorted block and receive $R + \theta$ and receive $-F$ otherwise. If they do not distort, they receive R with certainty. (This assumes that the attacker has full knowledge of the fact that they are the proposer before setting M_c). Thus, the proposer will try to attack if:

$$s(R + \theta) - (1 - s)F > R \implies s > \frac{R + F}{R + \theta + F} \tag{7.1}$$

From this, it can be seen that this mechanism is not robust to an attack if the attacker has a sufficient share (s) of the network.

We can compare this threshold to that typically considered for BFT networks. Attacks that may delay the confirmation of transactions are not possible in those networks if $s < \frac{1}{3}$. Notice here that the SR mechanism lowers this possibility if $R + F > \frac{\theta}{2}$. Thus, depending on the environment, the consensus protocol proposed here may be more secure than the usual

---

would be more complicated, but we conjecture that it will still hold given the results of Chen et al. (2022) that show that SR mechanisms are robust to some informational imperfections.

BFT protocol. Moreover, security can be enhanced by increasing $R + F$ rather than exogenously set as it is under the usual BFT consensus.[5]

The above calculations assume that an attack is considered by the attacker to be a once-off opportunity. This certainly is the case if, in the next round, a truthful consensus is reached and the opportunity for an attack is removed. However, when there are multiple nodes, the incentive compatibility conditions in our proposed mechanism need to be reformulated for the possibility that, should consensus not be reached with one pair, at least one additional round of the mechanism will result in a new pair of nodes. Gans and Holden (2022) show that the mechanism is vulnerable when $s > \frac{1}{2}$ drawing a similarity between these and 'majority' attacks in LCR blockchains.

7.3   Mechanism to Resolve Forks

Recall that a fork is a situation where two chains of equal length are confirmed to the protocol with a common ancestor but distinct blocks thereafter. When this occurs, nodes must decide which one to append new blocks to so that a new longest chain arises, which becomes the consensus chain. Because there is no voting protocol, LCR blockchains can have blocks confirmed very quickly, but because of the possibility of forks, consensus may not be final.

Forks are part of attacks to exploit LCR rules. An attacker privately works on a chain that nullifies a transaction in an already confirmed block that is part of the main chain. The transaction is removed so that the tokens remain in the attacker's account to be spent again. The goal is then to surface the private chain once it is as long as (or longer than) the current main chain; creating a fork.

---

[5] If we require nodes to send their messages *before* knowing who a proposer is, then this slightly changes the equation. Now the attacker only succeeds with probability s^2 but also has an additional cost in that their confirmer role automatically leads to a fine. Thus, an attacker's choice depends on:

$$s^2(R + \theta) - s(1 - s)F - (1 - s)F > R$$

and so the threshold becomes $s > \sqrt{\frac{R+F}{R+\theta+F}}$, which is an order of magnitude stronger (modulo, integer issues ignored here). In this case, the SR mechanism is more secure than the usual BFT consensus if $R + F > \frac{\theta}{8}$.

The attack is made more difficult if there are robust penalties to other nodes, from staking tokens to both chains in a fork. The LCR asks honest nodes to extend the longest chain. But, in reality, what would be preferable is if those nodes worked on the right chain. However, there is no real mechanism to determine which chain is the 'true' chain. That is where a mechanism can help.

The potential for a mechanism that might resolve blockchain forks arises because it is difficult to tell if they are accidental or deliberate. In the latter case, they are akin to an ownership dispute. A natural question is whether there is a mechanism that can quickly determine what the 'true' chain is.

A fork is only consequential if different nodes claim that said fork is the correct one. Thus, there are competing ownership claims about the 'true' chain. For simplicity, call the competing forks A and B. The information structure is such that the nodes claiming the fork A is the true one know they are honest. Nodes claiming that B is the true chain know they are not honest.

A Solomonic Mechanism

Now consider the following mechanism based on the Solomonic mechanism of the type we already discussed in Chapter 6 If a fork appears (without loss of generality, B) and is within x of the same number of blocks as fork A then the following mechanism is run between nodes that claim to hold a full record of the blockchain.

1. For each fork, the blocks subsequent to the last common parent are unpacked, and transactions are compared.
2. Valid messages that appear in both sets are immediately confirmed and at the minimum of the time-stamp between the two forks.[6] Other messages are collected and marked as disputed.
3. One node from each chain is selected at random (a for A and b for B). The node from the chain where a transaction does not appear is asked to confirm that the transaction is invalid. In this case,

---

[6] Valid messages will have different time-stamps but otherwise the same content. However, as they cover multiple blocks, there are some practical issues to resolve in comparing messages as being equivalent on each chain of the fork.

both are fined F, and they enter the dispute stage for each disputed transaction.

The **dispute stage** involves:

1. If the transaction appears in A and not in B, a is asked to assert the legitimacy of the transaction. If a asserts, then the transaction remains, and the fine is burned. If a does not assert, the transaction is discarded, and b has their fine refunded.
2. If the transaction appears in B and not in A, b is asked to assert the legitimacy of the transaction. If b asserts, then the transaction remains, and the fine is burned. If b does not assert, the transaction is discarded, and a has their fine refunded.

Note that there may be numerous transactions that appear in one chain and not the other. The procedure here, between the two selected nodes, would be conducted for each disputed transaction with the roles assigned depending on which chain the disputed transaction appears.

We now need to specify the preferences of each type of node. An honest node has an interest in preserving the true blockchain. That preference has a monetary equivalent value of H, and they have a disutility arising from another blockchain being built upon of D with $H > D$. By contrast, a dishonest node is only interested in having their preferred chain continue for which they receive a monetary equivalent value of θ.

Given this, the following can be shown:

Proposition 7 Dishonest forks do not arise in any subgame perfect equilibrium where at least one honest node is selected.

Proof Without loss in generality, suppose that A is the true blockchain and a dishonest fork, B, arises with a transaction omitted from a past block. If b confirms that the transaction is invalid, a and b are then selected and fined, and the dispute stage begins. In the dispute stage, a is asked to assert or not assert the legitimacy of the transaction. If a asserts as the transaction is valid and a is honest, a receives $H - F$, and b's payoff remains at $-F$. If a does not assert, a receives $D - F$ and b's payoff becomes θ. In this case, a has a preference to assert.

In the first round, anticipating this, b will decline to confirm the removal of the transaction as this will result in a fine of F for sure and no transaction being removed. Given that, it is not worthwhile attempting the attack in the first place.

Now suppose that there are no dishonest nodes and that, in this case, b is a node on a chain where the transaction does not appear (e.g., it may have been missed because of network issues). In this case, as b knows they are honest and that dishonest nodes only attempt to remove transactions and not place them, b chooses not to assert the transaction is invalid. Thus, the transaction will remain, and neither node will be fined.

The intuition for this result is simple. The mechanism is designed based on the notion that (1) nodes have information as to which chain they regard as truthful and which is not; and (2) the attacker is trying to have a past transaction/message removed from the blockchain. The mechanism gives the opportunity for a node that had staked on each chain to be matched and to confirm whether there is a dispute over a transaction that appears in one but not in the other. If there is an agreement, the transaction is confirmed or removed, as the case may be, and the mechanism ends. If there is a dispute, however, both nodes are fined which creates an incentive to avoid the dispute. The dispute stage then focuses on the node from the chain that asserts the transaction is valid. As an attack involves a dishonest node trying to remove a transaction, this node is presumptively honest and so is given control over the decision. Thus, the transaction remains. Given this is the outcome of the dispute, a dishonest node will not trigger the dispute (as they will be fined) nor create a dishonest fork because that is costly and will be unsuccessful.

Importantly, the mechanism takes into account the possibility that chains have arisen as a result of, say, network latency issues, rather than an attack. In this case, both nodes are honest. Given that nodes know that attacks involve the removal of transactions, when the mechanism is triggered by an accidental fork, the node from the chain without the transaction will agree their fork should be discarded if the transaction is, indeed, valid. Thus, the 'correct' fork persists.

Nonetheless, as the proposition qualifies, the mechanism does not prevent dishonest forks from arising when neither selected node is honest. This would require an attacker to distribute nodes across both forked chains. Gans and Holden (2022) consider this possibility and demonstrate that the above mechanism can actually create an incentive for attacking nodes to work on multiple forks. This may evade slashing if the attacker controls many different nodes. This is an issue if $s > \frac{1}{2}$. They show that an amended Solomonic mechanism can overcome this type of non-traditional double-spend attack.

While there will always be forks due to latency issues, there will never be a 'dishonest fork' on the equilibrium path of the game induced by these types of mechanisms. Indeed, provided that the honest nodes have a higher prior that the correct chain is, indeed, the correct chain—that is, priors are biased towards the truth—the mechanism here will still have them acting as if the correct chain is the truth and will still deter attempts to confirm a dishonest chain as the consensus chain.

7.4   Conclusion

Here we have seen how to construct revelation mechanisms to achieve consensus on blockchains under BFT and LCR. A fundamental pillar of a mechanism-design approach to blockchain consensus is the use of information. Consistent with the central dogma of mechanism design, the designer is not presumed to possess more information than that held by agents in the mechanism. This contrasts, however, with some uses of slashing in existing approaches to consensus. Yet, through careful design choices, the mechanisms we examine here make very efficient of the information that is held by existing nodes. We have also discussed the robustness of these mechanisms to multi-node attacks.

Mechanism-designed-based consensus protocols have important advantages over existing Proof of Stake protocols based on rules, and they are likely to be of practical use as an alternative to Proof of Work protocols. In particular, rather than just being designed to satisfy stringent requirements of finality and liveness (themselves often probabilistic), mechanisms have choice parameters (e.g., the reward, R, and fines, F) that themselves can manipulate trade-offs between finality and liveness at the margin depending on the environment and preferences of blockchain users. This expands the set of blockchain consensus options for participants.

Key insights from Chapter 7
- Blockchain consensus protocols specify rules designed to achieve consensus in a timely manner. However, those rules do not guarantee that consensus gives rise to truthful information in the sense that accurate information is confirmed to blocks and information is

116   J. GANS

not censored from blocks. Mechanisms can potentially be deployed
to resolve such issues.

- For Proof of Stake networks based on BFT, a simultaneous report
  mechanism can ensure that leaders propose blocks that accurately
  represent transactions in the mempool.
- For Proof of Stake networks based on LCR, a Solomonic mechanism
  can be used to resolve forks as an alternative to LCR consensus and
  to ensure that attacking chains do not become the consensus chain.

REFERENCES

Akbarpour, M., & Li, S. (2020). Credible auctions: A trilemma. *Econometrica*, *88*(2), 425–467.

Buterin, V. (2017). Minimal slashing conditions. *Medium*.

Chen, Y.-C., Holden, R., Kunimoto, T., Sun, Y., & Wilkening, T. (2022). Getting dynamic implementation to work. *Journal of Political Economy, 131*(2), 285–387.

Eliaz, K. (2002). Fault tolerant implementation. *The Review of Economic Studies, 69*(3), 589–610.

Gans, J. S., & Holden, R. (2022). *Mechanism design approaches to blockchain consensus*. Technical report, National Bureau of Economic Research.

Hurwicz, L. (1972). On informationally decentralized systems. *Decision and organization: A volume in Honor of J. Marschak*. North Holland Publishing Company.

Moore, J., & Repullo, R. (1988). Subgame perfect implementation. *Econometrica, 56*(6), 1191–1220.

REFERENCES

Abadi, J., & Brunnermeier, M. K. (2022). *Blockchain economics*. Technical report, Federal Reserve Bank of Philadelphia.

Aghion, P., Fudenberg, D., Holden, R., Kunimoto, T., & Tercieux, O. (2012). Subgame-perfect implementation under information perturbations. *The Quarterly Journal of Economics, 127*(4), 1843–1881.

Akbarpour, M., & Li, S. (2020). Credible auctions: A trilemma. *Econometrica, 88*(2), 425–467.

Aune, R. T., Krellenstein, A., O'Hara, M., & Slama, O. (2018). Footprints on a blockchain: Trading and information leakage in distributed ledgers. *The Journal of Trading, 13*(4), 49–57.

Ayres, I., & Talley, E. (1994). Solomonic bargaining: Dividing a legal entitlement to facilitate coasean trade. *Yale LJ, 104,* 1027.

Babbage, C. (1832). *On the economy of machinery and manufactures*. Taylor & Francis.

Back, A., et al. (2002). *Hashcash—A denial of service counter-measure*. Technical report.

Bakos, Y., & Halaburda, H. (2021). *Tradeoffs in permissioned vs permissionless blockchains: Trust and performance*. NYU Stern School of Business working paper.

Becker, G. S. (1968). Crime and punishment: An economic approach. In *The economic dimensions of crime* (pp. 13–68). Springer.

Biais, B., Bisiere, C., Bouvard, M., & Casamatta, C. (2019). The blockchain folk theorem. *The Review of Financial Studies, 32*(5), 1662–1715.

Brandenburger, A., & Steverson, K. (2021). *Using 'proof-of-presence' to coordinate*. Technical report.

Buchman, E. (2016). *Tendermint: Byzantine fault tolerance in the age of blockchains* [Ph.D. thesis].

Budish, E., Cramton, P., & Shim, J. (2015). The high-frequency trading arms race: Frequent batch auctions as a market design response. *The Quarterly Journal of Economics, 130*(4), 1547–1621.

Budish, E. B. (2022). *The economic limits of bitcoin and anonymous, decentralized trust on the blockchain.* University of Chicago, Becker Friedman Institute for Economics Working Paper No. 83.

Buterin, V. (2016, December 30). A proof of stake design philosophy. *Medium.* https://medium.com/@VitalikButerin/a-proof-of-stake-design-philosophy-506585978d51.

Buterin, V. (2017). Minimal slashing conditions. *Medium.*

Buterin, V. (2021, June). Proposer/block builder separation-friendly fee market designs.

Buterin, V., & Griffith, V. (2017). *Casper the friendly finality gadget.* arXiv preprint arXiv:1710.09437

Canidio, A., & Danos, V. (2022). *Commitment against front running attacks.* Technical report.

Catalini, C., & Gans, J. S. (2020). Some simple economics of the blockchain. *Communications of the ACM, 63*(7), 80–90.

Chen, Y.-C., Holden, R., Kunimoto, T., Sun, Y., & Wilkening, T. (2022). Getting dynamic implementation to work. *Journal of Political Economy, 131*(2), 285–387.

Chiu, J., & Koeppl, T. V. (2017). *The economics of cryptocurrencies–bitcoin and beyond.* Available at SSRN 3048124.

Cong, L. W., He, Z., & Li, J. (2021). Decentralized mining in centralized pools. *The Review of Financial Studies, 34*(3), 1191–1235.

Copeland, T. (2021, November). A DEX on cosmos is working on a way to prevent front running. *The Block.*

Daian, P., Goldfeder, S., Kell, T., Li, Y., Zhao, X., Bentov, I., Breidenbach, L., & Juels, A. (2019). *Flash boys 2.0: Frontrunning, transaction reordering, and consensus instability in decentralized exchanges.* arXiv preprint arXiv:1904.05234

De Vries, A. (2018). Bitcoin's growing energy problem. *Joule, 2*(5), 801–805.

Dolev, D., & Strong, H. R. (1983). Authenticated algorithms for byzantine agreement. *SIAM Journal on Computing, 12*(4), 656–666.

Dwork, C., & Naor, M. (1992). Pricing via processing or combatting junk mail. In *Annual international cryptology conference* (pp. 139–147). Springer.

Eliaz, K. (2002). Fault tolerant implementation. *The Review of Economic Studies, 69*(3), 589–610.

Eskandari, S., Moosavi, M., & Clark, J. (2019). *SoK: Transparent dishonesty: Front-running attacks on blockchain.* Springer.

Eyal, I., & Sirer, E. G. (2018). Majority is not enough: Bitcoin mining is vulnerable. *Communications of the ACM, 61*(7), 95–102.

Fama, E. F. (1980). Banking in the theory of finance. *Journal of Monetary Economics, 6*(1), 39–57.

Farrell, J., & Saloner, G. (1985). Standardization, compatibility, and innovation. *The RAND Journal of Economics, 16*(1), 70–83.

Ferreira Torres, C., Camino, R., et al. (2021, August 11–13). Frontrunner jones and the raiders of the dark forest: An empirical study of frontrunning on the ethereum blockchain. In *USENIX Security Symposium, Virtual*.

Fischer, M. J., Lynch, N. A., & Merritt, M. (1986). Easy impossibility proofs for distributed consensus problems. *Distributed Computing, 1*(1), 26–39.

Fischer, M. J., Lynch, N. A., & Paterson, M. S. (1985). Impossibility of distributed consensus with one faulty process. *Journal of the ACM (JACM), 32*(2), 374–382.

Gans, J. S. (2022). The fine print in smart contracts. In M. Corrales Compagnucci, M. Fenwick, & S. Wrbka (Eds.), *Smart contracts technological, business and legal perspectives* (Chapter 2). Hart Publishing.

Gans, J. S., & Gandal, N. (2021). Consensus mechanisms for the blockchain. In *The Palgrave Handbook of Technological Finance* (pp. 269–286). Springer.

Gans, J. S., & Halaburda, H. (2015). Some economics of private digital currency. In *Economic analysis of the digital economy*, 257–276. University of Chicago Press.

Gans, J. S., & Halaburda, H. (2023). *"Zero cost" double-spend attacks on permissionless blockchains*. Technical report.

Gans, J. S., & Holden, R. (2022a). *Mechanism design approaches to blockchain consensus*. Technical report, National Bureau of Economic Research.

Gans, J. S., & Holden, R. (2022b). *A solomonic solution to ownership disputes: Theory and applications*. Technical report, Mimeo.

Gans, J. S., & Holden, R. (2023). A solomonic solution to blockchain frontrunning. *American Economic Association: Papers and Proceedings, 113*, 248–252.

Haber, S., & Stornetta, W. S. (1990). How to time-stamp a digital document. In *Conference on the Theory and Application of Cryptography* (pp. 437–455). Springer.

Halaburda, H., Haeringer, G., Gans, J., & Gandal, N. (2022). The microeconomics of cryptocurrencies. *Journal of Economic Literature, 60*(3), 971–1013.

Halaburda, H., He, Z., & Li, J. (2021). *An economic model of consensus on distributed ledgers*. Technical report, National Bureau of Economic Research.

Halpern, J. Y., & Moses, Y. (1990). Knowledge and common knowledge in a distributed environment. *Journal of the ACM (JACM), 37*(3), 549–587.

Holden, R., & Malani, A. (2021). *Can blockchain solve the hold-up problem in contracts?* Cambridge University Press.

Holden, R., & Malani, A. (2021). *Can blockchain solve the hold-up problem in contracts? Elements in law.* Economics and Politics: Cambridge University Press.

Huberman, G., Leshno, J. D., & Moallemi, C. (2021). Monopoly without a monopolist: An economic analysis of the bitcoin payment system. *The Review of Economic Studies, 88*(6), 3011–3040.

Hurwicz, L. (1972). On informationally decentralized systems. *Decision and organization: A volume in Honor of J. Marschak.*

Keynes, J. M. (1937). The general theory of employment. *The Quarterly Journal of Economics, 51*(2), 209–223.

King, S., & Nadal, S. (2012, 19 August). *Ppcoin: Peer-to-peer crypto-currency with proof-of-stake.* Self-published paper (1).

Kocherlakota, N. R. (1998). Money is memory. *Journal of Economic Theory, 81*(2), 232–251.

Kwon, J. (2014). Tendermint: Consensus without mining. *Draft v. 0.6, fall 1*(11).

Leshno, J. D., & Strack, P. (2020). Bitcoin: An axiomatic approach and an impossibility theorem. *American Economic Review: Insights, 2*(3), 269–86.

Ma, J., Gans, J. S., & Tourky, R. (2018). *Market structure in bitcoin mining.* Technical report, National Bureau of Economic Research.

Malkhi, D., & Szalachowski, P. (2022). *Maximal extractable value (MEV) protection on a DAG.* arXiv preprint arXiv:2208.00940

Moore, J. (1992). Implementation, contracts, and renegotiation in environments with complete information. *Advances in Economic Theory, 1*, 182–282.

Moore, J., & Repullo, R. (1988). Subgame perfect implementation. *Econometrica, 56*(5), 1191–1220.

Moroz, D. J., Aronoff, D. J., Narula, N., & Parkes, D. C. (2020). *Double-spend counterattacks: Threat of retaliation in proof-of-work systems* arXiv preprint arXiv:2002.10736

Nakamoto, S. (2008). Bitcoin: A peer-to-peer electronic cash system. *Decentralized Business Review*, 21260.

Pease, M., Shostak, R., & Lamport, L. (1980). Reaching agreement in the presence of faults. *Journal of the ACM (JACM), 27*(2), 228–234.

Reinganum, J. F. (1989). The timing of innovation: Research, development, and diffusion. *Handbook of Industrial Organization, 1*, 849–908.

Robinson, D., & Konstantopoulos, G. (2020, August). Ethereum is a dark forest. *Medium.*

Roşu, I., & Saleh, F. (2021). Evolution of shares in a proof-of-stake cryptocurrency. *Management Science, 67*(2), 661–672.

Rubinstein, A. (1989). The electronic mail game: Strategic behavior under "almost common knowledge." *The American Economic Review, 79*(3), 385–391.

Saleh, F. (2021). Blockchain without waste: Proof-of-stake. *The Review of Financial Studies, 34*(3), 1156–1190.

Shapiro, C., & Stiglitz, J. E. (1984). Equilibrium unemployment as a worker discipline device. *The American Economic Review, 74*(3), 433–444.

Stankovic, S. (2021, August). What is MEV? Ethereum's invisible tax explained. *Cryptobriefing*.

Tullock, G. (2001). Efficient rent seeking. In *Efficient rent-seeking* (pp. 3–16). Springer.

Von Neumann, J., & Morgenstern, O. (1944). *Theory of games and economic behavior*. Princeton University Press.

INDEX